User-Centered Evaluation of Visual Analytics

Synthesis Lectures on Visualization

David Ebert, *Purdue University*
Niklas Elmqvist, *University of Maryland*

Synthesis Lectures on Visualization publishes 50- to 100-page publications on topics pertaining to scientific visualization, information visualization, and visual analytics. Potential topics include, but are not limited to: scientific, information, and medical visualization; visual analytics, applications of visualization and analysis; mathematical foundations of visualization and analytics; interaction, cognition, and perception related to visualization and analytics; data integration, analysis, and visualization; new applications of visualization and analysis; knowledge discovery management and representation; systems, and evaluation; distributed and collaborative visualization and analysis.

User-Centered Evaluation of Visual Analytics
Jean Scholtz

ISBN: 978-3-031-01477-2 print
ISBN: 978-3-031-02605-8 ebook

DOI 10.1007/978-3-031-02605-8

A Publication in the Springer series
SYNTHESIS LECTURES ON VISUALIZATION #9
Series Editors: David S. Ebert, Purdue University, Niklas Elmqvist, University of Maryland

Series ISSN 2159-516X Print   2159-5178 Electronic

User-Centered Evaluation of Visual Analytics

Jean Scholtz

Pacific Northwest National Laboratory (PNNL)

SYNTHESIS LECTURES ON VISUALIZATION #9

ABSTRACT

Visual analytics has come a long way since its inception in 2005. The amount of data in the world today has increased significantly and experts in many domains are struggling to make sense of their data. Visual analytics is helping them conduct their analyses. While software developers have worked for many years to develop software that helps users do their tasks, this task is becoming more and more onerous, as understanding the needs and data used by expert users requires more than some simple usability testing during the development process. The need for a user-centered evaluation process was envisioned in *Illuminating the Path*, the seminal work on visual analytics by James Thomas and Kristin Cook in 2005. We have learned over the intervening years that not only will user-centered evaluation help software developers to turn out products that have more utility, the evaluation efforts can also help point out the direction for future research efforts.

This book describes the efforts that go into analysis, including critical thinking, sensemaking, and various analytics techniques learned from the intelligence community. Support for these components is needed in order to provide the most utility for the expert users. There are a good number of techniques for evaluating software that hasbeen developed within the human-computer interaction (HCI) community. While some of these techniques can be used as is, others require modifications. These too are described in the book. An essential point to stress is that the users of the domains for which visual analytics tools are being designed need to be involved in the process. The work they do and the obstacles in their current processes need to be understood in order to determine both the types of evaluations needed and the metrics to use in these evaluations. At this point in time, very few published efforts describe more than informal evaluations. The purpose of this book is to help readers understand the need for more user-centered evaluations to drive both better-designed products and to define areas for future research. Hopefully readers will view this work as an exciting and creative effort and will join the community involved in these efforts.

KEYWORDS

visual analytics, user-centered evaluations, sensemaking, analytic techniques, human-computer interaction, metrics

Contents

Acknowledgments

I appreciate the support of my colleagues at the Pacific Northwest National Laboratory who have not only helped with this work but also have supported user-centered design and evaluation work on a wide variety of projects. A special thank you to Niklas Elmquivst for encouraging me to produce a work specifically addressing the evaluation of visual analyst systems from the user's point of view.

CHAPTER 1

Introduction

Visual analytics systems are becoming very popular. More domains now use interactive visualizations to analyze the ever-increasing amount and heterogeneity of data. More novel visualizations are being developed for more tasks and users. We need to ensure that these systems can be evaluated to determine that they are both useful and usable. A user-centered evaluation for visual analytics needs to be developed for these systems. While many of the typical human-computer interaction (HCI) evaluation methodologies can be applied as is, others will need modification. Additionally, new functionality in visual analytics systems needs new evaluation methodologies.

There is a difference between usability evaluations and user-centered evaluations. Usability looks at the efficiency, effectiveness, and user satisfaction of users carrying out tasks with software applications. User-centered evaluation looks more specifically at the utility provided to the users by the software. This is reflected in the evaluations done and in the metrics used. In the visual analytics domain this is very challenging as users are most likely experts in a particular domain, the tasks they do are often not well defined, the software they use needs to support large amounts of different kinds of data, and often the tasks last for months. These difficulties are discussed more in the section on User-centered Evaluation.

Our goal is to provide a discussion of user-centered evaluation practices for visual analytics, including existing practices that can be carried out and new methodologies and metrics that need to be developed and agreed upon by the visual analytics community. The material provided here should be of use for both researchers and practitioners in the field of visual analytics. Researchers and practitioners in HCI who are interested in visual analytics will find this information useful, including a discussion on changes that need to be made to current HCI practices to make them more suitable to visual analytics. A history of analysis and analysis techniques and problems is provided as well as an introduction to user-centered evaluation and various evaluation techniques for readers from different disciplines. The understanding of these techniques is imperative if we wish to support analysis in the visual analytics software we develop.

Currently, the evaluations that are conducted and published for visual analytics software are very informal and consist mainly of comments from users or potential users. Our goal is to help researchers in visual analytics to conduct more formal user-centered evaluations. While these are time-consuming and expensive to carryout, the outcomes of these studies will have a defining impact on the field of visual analytics and help point the direction for future features and visualizations to incorporate.

While many researchers view work in user-centered evaluation as a less-than-exciting area to work, the opposite is true. First of all, the goal of user-centered evaluation is to help visual analytics software developers, researchers, and designers improve their solutions and discover creative ways to better accommodate their users. Working with the users is extremely rewarding as well. While we use the term "users" in almost all situations there is a wide variety of users that all need to be accommodated. Moreover, the domains that use visual analytics are varied and expanding. Just understanding the complexities of a number of these domains is exciting. Researchers are trying out different visualizations and interactions as well. And, of course, the size and variety of data are expanding rapidly. User-centered evaluation in this context is rapidly changing. There are no standard processes and metrics and thus those of us working on user-centered evaluation must be creative in our work with both the users and with the researchers and developers. We hope that this review of visual analytics and possibilities for starting work in user-centered evaluation in this area helps to point out possibilities and to generate excitement for more research in this area.

CHAPTER 2

Analysis

For many readers, the term "analysis" automatically combines with "intelligence." Visual analytics are used in many fields, such as banking, transportation planning, climate science, pandemics, and emergency response, as well as in intelligence analysis. However, many of the techniques and studies referenced are from the field of intelligence analysis. This field created much of the science of analysis during the cold war and these techniques were adopted by corporations basically unchanged. In order to understand what needs to be done to evaluate the use of visual analytics tools, it is necessary for researchers and practitioners to understand the evolution of the practice of analysis including the tools and metrics used in that field. In the rest of this section, the history of intelligence analysis is provided along with a discussion of the various techniques used by intelligence analysts as those have been adopted in many other fields as well. We also include a discussion of the standards that intelligence analysts apply to their thinking to ensure that they have adequately and fairly covered all the different aspects of an issue.

2.1 HISTORY OF INTELLIGENCE ANALYSIS IN THE UNITED STATES

In the United States, intelligence has been collected and disseminated since the American Revolution (CIA, Notes from our Attic, 2017). Prior to World War II, intelligence analysis in the U.S. was done separately by the U.S. Department of State (mostly during official business) and the armed services. In 1941, a Coordinator of Information (COI), William Donovan, was appointed by President Franklin Roosevelt to coordinate intelligence between the different agencies. The COI was to collect all intelligence from the different agencies doing the collection, analyze the information, and distribute it to the President and to various agencies as needed (CIA Library/Intelligence History, 2017). The COI was split in 1941, with the portion responsible for public radio broadcasting becoming the Foreign Information Service and the rest of the COI became the Office of Strategic Services (OSS). OSS staff learned much from closely working with the British intelligence service. Research and Analysis (R&A), an OSS branch, was responsible for inventing non-departmental strategic analysis. After World War II, the OSS was disbanded but the R&A group continued. In 1947, the need for a centralized intelligence agency was recognized and President Truman established the U.S. Central Intelligence Agency (CIA).

2.2   CHANGES IN INTELLIGENCE ANALYSIS

Intelligence analysis during the Cold War was based on the process of "gathering, analyzing, estimating and disseminating information" (Baumard, 1994). It was this process that was incorporated into corporations unchanged. Since this time, both the intelligence community and corporations have evolved from a fairly static environment to a highly dynamic environment and have had to revise their analysis process (Moore, 2011).

2.3   SHERMAN KENT AND THE DISCIPLINE OF INTELLIGENCE ANALYSIS

Sherman Kent is often referred to as the "father of intelligence analysis." He worked in the OSS and later in the CIA during World War II and the Cold War era and pioneered many intelligence techniques. He wrote *Strategic Intelligence for American World Policy* in 1949. In 2000, the CIA opened The Sherman Kent School for Intelligence Analysis.

An essay by Kent (1955) notes that during WWII the U.S. was almost entirely dependent on the British intelligence service. However, he states that by 1955, the intelligence community had gone from a profession to almost a discipline and hence needed a body of literature that dealt with first principles in order to reach full maturity. Scholars and others interested in the literature produced by the intelligence community can find unclassified papers from 1992 to the current day at the CIA Library Center for the Study of Intelligence (2017).

2.4   CRITICAL THINKING

Critical thinking is an essential component of intelligence analysis. Moore's (2007) definition of critical thinking is "that mode of thinking—about any subject, content, or problems—in which the [solitary] thinker improves the quality of his or her thinking by skillfully taking charge of the structures inherent in thinking and imposing intellectual standards upon them."

Paul and Elder (2004) also outline eight elements of reasoning. These include the purpose of the analysis, the question being addressed, the evidence, the inference and interpretations made, the assumptions made, the implications and consequences of those, and the points of view that are represented. In addition, a set of nine intellectual standards for evaluating the quality of one's thinking are given by Paul and Elder. These include assessing the logic, clarity, accuracy, precision, depth, significance, and fairness of the reasoning. While these questions and standards may seem trivial, it is often helpful to consider these when evaluating the thinking that has gone into an intelligence product. Again, the issue of whether the important aspects been covered and the analysis has been done at the highest standards should be supported by a software tool for visual analysis. Standards similar to these are used today to critique analytics products in the intelligence community. Analysts are not always right but often errors can be caught and corrected by critical thinking and applying the standards for evaluating the quality of the analysis. We will not discuss deception here but careful critical thinking may help understand if deception is occurring.

Producing intelligence, in any field, demands that the producer use critical thinking skills in first thinking through the problem, obtaining facts that are relevant to the question, making sure that all points of view have been considered, that any assumptions that have been made are clearly stated, and that the inferences and interpretations made are clear and logical. While technology can certainly aid analysts in their thinking and help them find information more easily, analysts still need to monitor the process they are using. Moreover, those of us who design and develop systems to help analysts must realize that the software needs to work *for* the analysts, rather than be worked *by* analysts. Every cycle we take away from the analyst to deal with software decisions is one less cycle for her to concentrate on the analytic process. This notion as well as the use of the components above along with the intellectual standards should be a guiding principle of the user-centered evaluation of visual analytics software.

CHAPTER 3

Analytic Methods

In addition to critical thinking, which should be employed in any type of analysis, various analysis methods are used by analysts (Jones, 1998). Humans too often let emotions affect their rationality when thinking through problems or questions. Often we are content to settle for the first solution found. Other times, one element of a solution is acceptable so the fact that the rest of the solution is not suitable is ignored. In addition to critical thinking, a number of specific techniques can be used to structure our thinking and ensure that a number of objectives are considered. One of the most common techniques is Analysis of Competing Hypotheses (ACH) (Heuer, 1999). Using this method, the analyst considers a set of hypotheses, and then looks for evidence that supports or refutes the different hypotheses. Different mathematical methods can be used to score the evidence and determine which hypothesis has the strongest supporting evidence. Other techniques such as the Devil's Advocate, the use of timelines, and decision/event trees are discussed in Jones (1998).

3.1 SENSEMAKING

Sensemaking is the process that analysts use in their work to give meaning to experience. An analyst typically takes a problem or request for information and sets about looking for information, saving pertinent parts, constructing hypotheses, finding evidence for those hypotheses, and so on, until she feels she has a good recommendation or assessment of the situation. Depending on the information requested or problem at hand, this process can take from several hours to many months.

In general, much of the work in visual analytics concerns "wicked problems." The term "wicked problem" is attributed to Rittel and defined as "a class of social system problems which are ill-formulated, where the information is confusing, where there are many clients and decision makers with conflicting values, and where the ramifications in the whole system are thoroughly confusing (Buchanan, 1992). Because of complex interdependencies, the effort to solve one aspect of a wicked problem may reveal or create other problems. In fact many of the tasks given to analysts fall into the category of a wicked problem. In addition to the characteristics of wicked problems, the issue of deception may come into play as well.

Analysts not only evaluate the information they find but also evaluate the process they are using to make sense of information. The activities carried out in sensemaking are planning, foraging, marshaling, understanding, and communicating. At each step, analysts are questioning and assessing. Pirolli and Card (2005) express the sensemaking loop in a slightly different fashion in Figure 3.1. This figure depicts the steps that will be carried out in a software tool. The planning step is not shown as this is most likely carried out mentally and most likely not documented. The foraging and marshaling steps are show as searching for information/relations/evident and pulling this information in the shoebox and evident files. The loops between the schema and hypotheses boxes represent the understanding phase while the presentation box is preparing the communication portion. Essentially, Pirolli and Card have shown sensemaking as distinct from the foraging loop but connected to it. They also add in a reality/policy loop that connects the overall process to the overall task of the analyst.

Klein et al. (2007) propose six steps of the data-frame sensemaking process: (1) sensemaking is the process of fitting data into a frame and fitting a frame around the data; (2) therefore, the "data" are inferred, using the frame, rather than being perceptual primitives; (3) the frame is inferred from a few key anchors; (4) the inferences used in sensemaking rely on abductive reasoning as well as logical deduction; (5) sensemaking usually ceases when the data and frame are brought into congruence; (6) experts reason the same way as novices but have a richer repertoire of frames. Sensemaking is used to achieve both a functional understanding—what to do in a situation—and an abstract understanding. People primarily rely on just-in-time mental models. Sensemaking takes different forms, each with its own dynamics.

Russell et al. (1993) find that analysts, to make sense of information, use the following steps.

- *Search for representations*. The sensemaker creates representations to capture important regularities in a way that supports the use of the instantiated representation. This is the **generation** loop.

- *Instantiate representations*. The sensemaker identifies information of interest, encoding it in an appropriate representation. Instantiated schemas are called **encodons** and are created in the **data coverage** loop.

- *Shift representations*. Representation shifts during sensemaking are intended to reduce the cost of the task operations. Forcing a change to the representation in this way is a bottom-up or data-driven process. **Residue** refers to ill-fitting or missing data and unused representations. The **representational shift** loop is guided by the discovery of residue. When there are relevant data without a place in the representation, the schemas can be expanded. When data do not fit the established categories, the original schema categories may need to be merged or split, or new categories may be added. Thus, sensemaking iterates between the top-down representation instantiation and bottom-up representation search processes.

- *Consume encodons*. The sensemaker then uses the encodons in some task-specific information processing step. In sensemaking, schemas provide top-down or goal-directed guidance. They prescribe what to look for in the data, what questions to ask, and how the answers are to be organized.

- But representation search is not simply top-down. If there were no surprises in creating encodons, sensemaking would be trivial—merely define the schemas and then instantiate them.

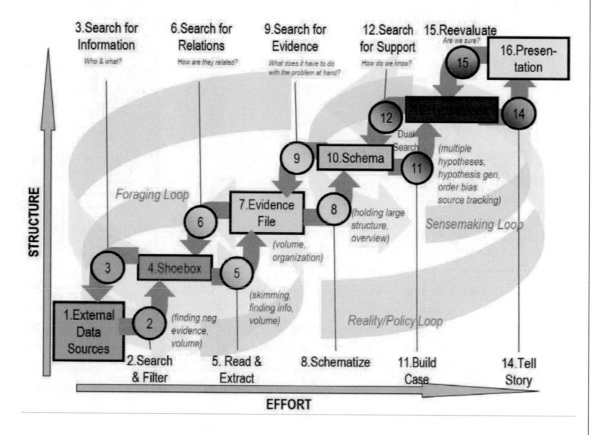

Figure 3.1: The sensemaking loop. From Pirolli and Card (2005). Used with permission.

Bell (1997) developed a tool to be used in teaching science to middle school students that helps them with sensemaking. Figure 3.2 shows an argument constructed about how far light goes by a team of students. The students are asked to put in evidence and the strength of that evidence to support their claims.

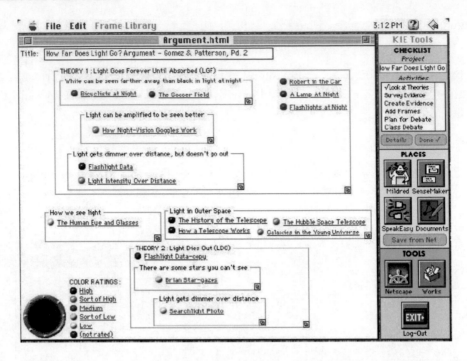

Figure 3.2: A tool used for middle school students to detail their sensemaking in science questions (Bell, 1997). Used with permission.

Sensemaking is seldom straightforward. Schemas must be revised when there are surprises creating encodons or as new task requirements come to light. All three different approaches have in common the existence of an iterative sensemaking loop and the notion that analysts go about this process in many different ways. However, whatever system the analyst uses needs to provide support for sensemaking activities. That is, support is needed not just for interacting with visualizations and finding information but also for the sensemaking process. System support for sensemaking will be discussed further in the evaluation section.

CHAPTER 4

What is Visual Analytics and Why is it Needed?

Support for sensemaking has been around for some time. Intelligence analysts often used spreadsheets to collect their facts and evidence, often adding new columns or augmenting information in columns as needed as information changed in format. However, the September 11, 2001 attacks changed the U.S. focus on analysis. It became apparent that we need some way to process more data and to supply tools to the analysts to help them analyze this data. The U.S. Department of Homeland Security chartered the National Visualization and Analytics Center TM at the Pacific Northwest National Laboratory (PNNL) to help. PNNL worked with researchers to understand what was needed to develop capabilities that would aid in preventing such attacks in the future. As a result of this work, Thomas and Cook (2005) published an agenda for a long-term visual analytics research and development program. Many disciplines were involved in the work leading up to the grand challenge below:

> [Support] the analysis of overwhelming amount of disparate, conflicting, and dynamic information to identify and prevent emerging threats, protect our borders, and respond in the event of an attack or other disaster.

The group identified visual analytics development as a key to achieving this grand challenge. Thomas and Cook (2005) define visual analytics as "the science of analytical reasoning facilitated by interactive visual interfaces." They lay out the following challenges for visual analytics research and development:

- the science of analytical reasoning;

- visual representations and interaction technologies;

- data presentations and transformations;

- presentation, production, and dissemination;

- moving research into practice; and

- positioning for enduring success.

The focus in the following sections is on evaluation of visual analytic systems which is discussed in the Thomas and Cook (2005) chapter on "Moving Research into Practice." As the technologies to support visual analytics are developed, evaluating these efforts helps both to move them into practice and to help researchers improve their work. The evaluations of visual analytics software are more complex than typical HCI evaluations. The preceding sections on analysis should help in understanding some of the issues that analysts face and the help that visual analytics systems strive to provide. The following sections discuss typical HCI evaluation methods as well as new evaluations or changes to typical HCI evaluations needed for visual analytics software.

While visual analytics work was first defined in the U.S. others quickly recognized this need and published challenges and needs as well (Keim et al., 2006; Keim et al., 2008a).

4.1   EXAMPLES OF VISUAL ANALYTICS TOOLS

Palantir started their work in 2004 looking at ways to support analysts. In 2017, Palantir supports two visual analytics tools, Gotham and Metropolis. Gotham is able to support both structured data, such as spreadsheets and log files, as well as unstructured data, including emails, videos, and documents. This data is imported into the tool and the user can see relationships between people, places, and objects. As new data appears, it is pulled into the system so that users are always working with the most up to date data. Security is built into the application so that only those authorized are able to view particular data. Gotham provides visualizations such as graphs and maps to support semantic, geospatial, temporal, and text analysis. Users can tag data to use in other Gotham applications. In addition, Gotham provides a mobile version so that it can be used in the field to support work in the field, such as when it was used to support disaster relief during Hurricane Sandy (https://www.palantir.com/products/; accessed August 2017).

Palintir's Metropolis tool is built to support such activities as analyzing network traffic flow and financial activity patters. Metropolis takes large amounts of quantitative data and builds abstractions of these for users to interact with. Users have access to models that can be used to build more complex models and to provide fast computations.

The Tableau Desktop (https://www.tableau.com/products/desktop; accessed August 2017) helps users interact with massive amounts of data on their desktops, building new visualizations in short amounts of time to understand the data better. It supports collaboration so that teams in an organization can work together. The visualizations help the users see trends and outliers quickly. Collaborative visual analytics systems are discussed in the section on Trends in Visual Analytics. Another interesting feature in Tableau is that of storytelling. Figures 4.1, 4.2, and 4.3 represent a series of visualizations that convey information to the end user about energy consumption in the U.S., the states that use the most, the providers of this energy, and the ownership of the providers, the same data but different visualizations.

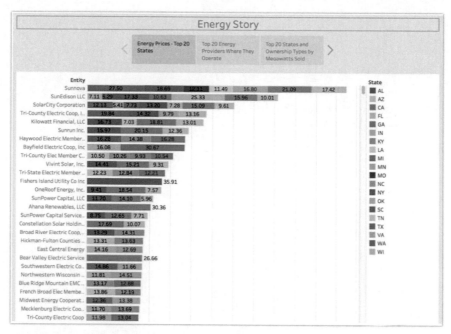

Figure 4.1: Energy process for the states that consume the most energy. Courtesy of Heywood Drummond, Tableau Software.

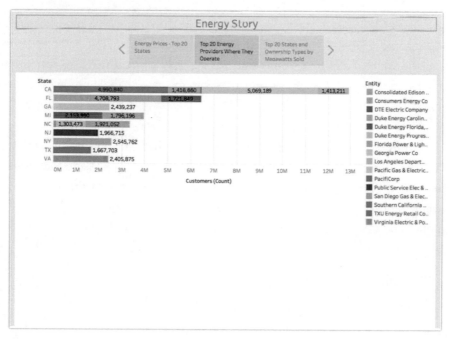

Figure 4.2: The top providers and the states where they operate. Courtesy of Heywood Drummond, Tableau Software.

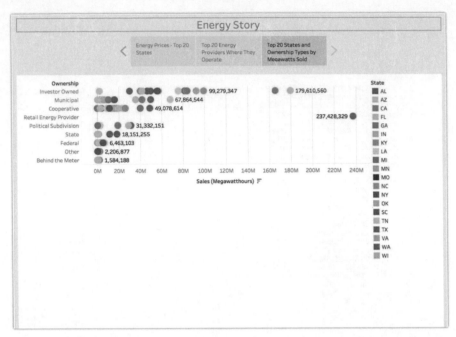

Figure 4.3: The ownership of the various energy providers by amounts sold. Courtesy of Heywood Drummond, Tableau Software.

Wright et al. (2006) developed a tool that provided the analyst a place to save notes and organize her analysis. Figure 4.4 shows their sandbox where the analyst has placed information and shown relationships discovered. While this was an early effort it needs to scale up to massive amounts of data. The key objective of this project was to increase analytic productivity and increase analytic quality.

The idea in these products as well as many others is that users should be working with the data, not with the user interface. The challenge in visual analytics is to ensure that users maximize the time they spend thinking about and working with the data and that the application supports what they want to do as seamlessly as possible. User-centered evaluation's goal is to use the evaluation techniques and metrics to ensure that visual analytic tools are designed and implemented to support this interaction.

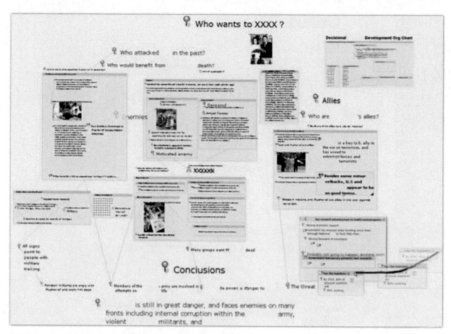

Figure 4.4: The Sandbox from the TRIST application developed in 2006 (Wright et. al, 2006). Used with permission.

CHAPTER 5

User-Centered Evaluation

Visual analytics systems are not only complex to design and implement, they also present multiple challenges in evaluation. While it is, of course, necessary to evaluate software to ensure that it functions according to specifications, this chapter is concerned with evaluating the software from the standpoint of how well it serves the user population. ISO 9241-11 defines usability as "the effectiveness, efficiency and satisfaction with which specified users achieve specified goals in particular environments." While evaluating the usability of the software is important, many additional and important questions need to be answered as well especially in evaluating the utility of new software. Usability issues need to be solved for specific tasks, but the following additional questions go beyond the usability of the software. They include questions of how well the software fits into processes in the user's workplace, if the user can accomplish her tasks more easily with the software, how difficult it is to import the users' data, if the visualizations are easily interpretable, how well the visualizations fit into their reasoning process, if the users can use their favorite analytic methods using this software, and if they are able to gain insights for their analysis.

As can be seen from the questions above, user-centered evaluation is a complex process that needs to occur in parallel with the design and development of the system. If the evaluation is done after the software has been fully completed, it may be extremely costly to fix any serious issues. We will discuss at what points in the software development cycle the evaluation techniques described in this chapter may be carried out. The earlier that mismatches between the software design and/ or implementation and the users' intended process are discovered the less expensive they are to fix.

Additionally, as stated earlier, components of the software should be evaluated prior to developing and evaluating the entire system. If the entire system is evaluated late in the development stage, it may be difficult to determine exactly what the problem is. If significant pieces of the system are evaluated early in the development cycle, then problems can fixed before those parts are integrated into the system. For example, users' reactions to visualizations can be obtained by giving users a task and determining if they can obtain the answer from the visualization. If interactions with the visualizations are to be examined, users could be shown "before and after" pictures of an interaction to gauge if the results are easy to understand.

Many visual analytics systems will be used daily, so even the evaluations we conduct during design and implementation may not uncover issues that arise with daily use. Field studies and the use of diaries or log files to view usage are some ways to ensure that the transition to actual daily use goes smoothly.

In discussing the various types of evaluation, it is also be necessary to talk about metrics and to understand the types of participants needed for various evaluation efforts. The actual users of visual analytics systems are often busy professionals who have little time to spend in evaluation sessions. At times, it may be satisfactory to find substitutes, but at other times, it may be necessary to get the actual users to participate or find appropriate surrogates. For example, visualizations that depict the spread of diseases could be tested with medical students in that field.

It is useful to look at a few abstractions to determine how evaluation efforts in one part of visual analytics software impact decisions made in other parts of the system. The Nested Blocks model (Munzner, 2009) shows the view of the different abstraction levels of a visual analytics system in Figure 5.1.

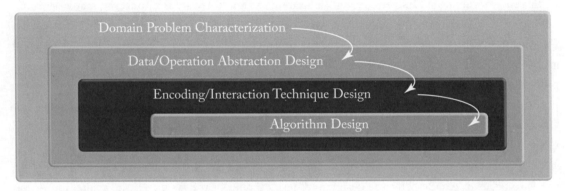

Figure 5.1: The Nested Blocks model of visual analytics systems. A nested model for visualization design and validation. Based on T. Munzner (2009).

In Figure 5.2, Munzner (2009) shows these abstraction levels along with the questions that need to be validated. Munzner uses the term "validation" to refer to both whether the right system, visualization, algorithm is used and if that system, visualization, algorithm is implemented correctly. The questions to be investigated are shown as threats. At the outer level, for example, is the threat that you are solving the wrong problem for the user. Once the system is implemented the evaluation proceeds to a different type of methodology for evaluation.

It is also necessary to discuss the dependencies between various parts of the visual analytics system. In later research, Meyer et. al. (2015) show the dependencies between the blocks. This version of the nested blocks is shown in Figure 5.3. The term "blocks" is used to mean an outcome of a design at any level in the system. A new visualization for data would be a block in the abstraction level. A way to render a specific visualization on a mobile phone would be a block at the algorithm level. The givens and what will be designed need to be considered. First the domain must be identified—i.e., identify the users' situation. The next block is the abstraction block, which comprises tasks and data. Tasks are identified based on the domain block but the data block is designed. The original data is identified and then transformed into the representation that will be used within the system. The technique block consists of visual encodings and interactions, both of which are designed. Algorithm blocks are also designed and focus on how to draw on the screen what the technique blocks are requesting. Guidelines are used for designs within the blocks. Any constraints imposed by a higher block on the block below need to be identified. Guidelines also exist for the transition between blocks. It is easy to see how decisions made in any one block affect the blocks above and below. This imposes restrictions on both the design process and the evaluation process as the evaluation may suggest changes between or within blocks.

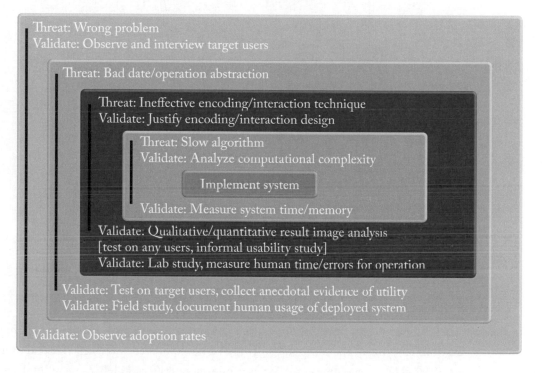

Figure 5.2: The questions to validate at each level in the system Munzner, T. (2009). A nested model for visualization design and validation. Based on T. Munzner (2009).

The typical HCI evaluation methods will be discussed first, followed by a discussion of how these methods can be used to evaluate visual analytics system and what, if any, modifications or additions are needed. Finally, we will revisit the nested blocks model and discuss possible sequences for user-centered evaluations.

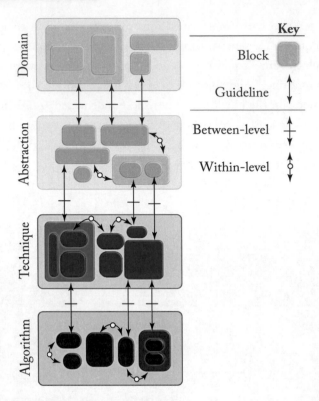

Figure 5.3: The nested blocks model. Based on Meyer et al. (2015).

5.1   TYPICAL HCI EVALUATION METHODS

When the first desktop computers were developed, the field of HCI emerged. People in this field felt it was important to ensure that the computer software being developed was usable for end users. Hence, a number of guidelines were developed for user interfaces. Early guidelines were published by Smith (1981), Smith and Aucella (1983), and Smith and Mosier (1984), with each iteration detailing more guidelines to accommodate the increasing capabilities of the user interfaces. Guidelines comprised a set of recommendations on how to design user interfaces that would be "user friendly." Shneiderman's early book (1987) presents information for designers to use in developing user interfaces. Following the development of a user interface, developers frequently conducted usability studies in a laboratory, watching various users attempt some assigned tasks with the software. As user interfaces became increasingly complex, so did the guidelines for their development. As it was no long feasible to check software to determine if the appropriate guidelines had been followed, the concept of heuristics and heuristic evaluations were developed. In addition, laboratory studies were only able to view the users' interactions with a specific piece of software, rather than see how it was used in their real-world context. Therefore, field studies were born. The sections below discuss each of these techniques for evaluation in typical desktop applications. While these are all techniques that could be used for evaluating visual analytics systems, the following section discusses changes and enhancements needed to adequately conduct these evaluations.

5.1.1   GUIDELINES

When the first personal computers were developed, various software organizations published guidelines used for developing their software. Guidelines have evolved to cover different applications but are still very evident. For example, Apple guidelines exist for iOS, macOS, watch OS, and tvOS; these guidelines are for developers designing applications for iPhone, Macintosh computers, Apple watches, and Apple TV, respectively. The goal of guidelines is to ensure that commands and icons used in various applications look alike and function in the same way so that once a user has learned the basics, she will have little trouble using a new application.

Likewise, Microsoft has similar guidelines. The guidelines for applications for Microsoft Windows (Microsoft Design Guidelines) separate the guidelines into those for controls, commands, text, messages, interaction, windows, visuals, experiences, and Windows environment. Controls, for example, provide guidelines for drop-down lists, combo boxes, group boxes, notifications, search boxes, sliders, tabs, tooltips, and info tips to name a subset.

The U.S. Department of Health and Human Services (HHS Standards and Guidelines) maintains a list of standards and guidelines for developing websites that are both usable and accessible. The guidelines have an associated strength of evidence and a relative importance assigned by a committee of usability researchers and web developers. To determine the relative importance of the guidelines, 16 external reviewers (8 website designers and 8 usability experts) looked at the guidelines and assigned a rating based on how important that guideline is to the success of a website from 1–5. The strength of evidence for each guideline was assigned by 8 usability researchers, practitioners, and authors (all with doctorates in an HCI-related field and experienced in HCI). The guidelines with a low rating for importance were eliminated from the library.

Academic studies are frequently conducted in psychology laboratories or HCI laboratories and focus primary on cognitive and visual aspects that provide guidance to user interface designers about design decisions that make the software more usable. Fitts' law is an example of this. Studies run by Fitts (1954) provided a model that predicts the time it takes a human to move the cursor to a particular element on the screen. While the actual model was developed for movement in the physical world, the model holds in the digital environment as well. This law was very useful in the layout of spatial elements on the computer screen to minimize the time it took users to select frequently used display elements.

One guideline is based on human limitations for information processing (Miller, 1956) and found that human can recall "seven plus or minus two" digits. Lindley (1966) found that this restriction could be overcome if one could break the numbers into chunks and then remember them as chunks rather than individual numbers. Hence, ten-digit phone numbers are often written (and remembered) as area code + local prefix + four-digit number.

Another example of a guideline addresses the order of items in a list. The guideline notes that lists should be arranged to facilitate the user's task. If there is no such order, then alphabetical or numerical order should be selected (HHS Standards and Guidelines for Order of Elements).

While guidelines are typically considered in designing software, they can also be used for evaluating software applications. Once a designer has a proposed design for an application, then an evaluator who is familiar with the guidelines can check to determine if the designer has adhered to the guidelines. Note that both the visual component of an object in a user interface and the functionality of the object are both covered by the guidelines. Following the guidelines and checking that the guidelines are not violated would be straightforward except that the guidelines are extremely numerous and the software world is forever creating new and better software that many times does not follow the conventional guidelines. Although designers have a good sense of the guidelines and do apply them in their work, specific guideline evaluation is rarely, if ever, carried out today. Therefore, heuristics and the technique of heuristic evaluation were developed and are discussed in the following section.

5.1.2   HEURISTICS

The heuristic evaluation technique was developed by Nielsen and Molich (1990). Heuristics are sets of guidelines that can be grouped under various headings or "rules of thumb" as they are sometimes called. Heuristics are at a much higher level of abstraction than guidelines. Nielsen and Molich developed these heuristics by taking a known set of usability problems and noting the various guidelines that were violated and as a result created these problems (Nielsen and Molich, 1990). They created a set of ten higher-level rules of thumb or abstractions that described a number of specific guidelines. While guidelines are often specific to a particular operating system, heuristics are applicable to any operating system. These heuristics can be used to examine software for violations that suggest potential user problems. These heuristics have since been modified to make them more usable for human-computer evaluations (Nielsen, 1994a). The following are the ten heuristics published on the Nielsen Norman Group website (Nielsen Norman Group Website).

Visibility of system status

The system should always keep users informed about what is going on, through appropriate feedback within reasonable time.

Match between system and the real world

The system should speak the users' language, with words, phrases, and concepts familiar to the user, rather than system-oriented terms. Follow real-world conventions, making information appear in a natural and logical order.

User control and freedom

Users often choose system functions by mistake and will need a clearly marked "emergency exit" to leave the unwanted state without having to go through an extended dialogue. Support undo and redo.

Consistency and standards

Users should not have to wonder whether different words, situations, or actions mean the same thing. Follow platform conventions.

Error prevention

Even better than good error messages is a careful design which prevents a problem from occurring in the first place. Either eliminate error-prone conditions or check for them and present users with a confirmation option before they commit to the action.

Recognition rather than recall

Minimize the user's memory load by making objects, actions, and options visible. The user should not have to remember information from one part of the dialog to another. Instructions for use of the system should be visible or easily retrievable whenever appropriate.

Flexibility and efficiency of use

Accelerators—unseen by the novice user—may often speed up the interaction for the expert user such that the system can cater to both inexperienced and experienced users. Allow users to tailor frequent actions.

Aesthetic and minimalist design

Dialogs should not contain information which is irrelevant or rarely needed. Every extra unit of information in a dialog competes with the relevant units of information and diminishes their relative visibility.

Help users recognize, diagnose, and recover from errors

Error messages should be expressed in plain language (no codes), precisely indicate the problem, and constructively suggest a solution.

Help and documentation

Even though it is better if the system can be used without documentation, it may be necessary to provide help and documentation. Any such information should be easy to search, focused on the user's task, list concrete steps to be carried out, and not be too large.

Heuristic evaluations are useful tools as they can be done very early in the design process, beginning with specification documents or early prototypes. They are most often accomplished by experts in usability. It is useful to have more than one expert do a heuristic review, with five being the optimal number, as more experts will determine more usability issues. One issue with heuristic evaluations is that experts are often unable to determine the severity of the various usability problems uncovered (Nielsen, 1992).

Heuristic reviews have been developed for domains in HCI including websites, video games, ambient displays, and computer-supported cooperative tools among others. However, at this time, no commonly accepted heuristics are available for information visualization or for visual analytics systems although Forsell and Johansson (2010) documented ten potential heuristics.

5.1.3   USABILITY EVALUATIONS

Usability evaluation was originally of two types: formative and summative. A formative evaluation is conducted during the product design to test and compare several approaches to the design or to determine if a specific design would work for users (Dumas and Redish, 1999). A summative usability evaluation is often conducted before product delivery and evaluates the product using measures of effectiveness, efficiency, and user satisfaction (Albert and Tullis, 2013).

Typically, formative usability sessions are conducted with a small number of participants and are informal evaluations that might show a participant several design mock-ups and ask which participants preferred. Each participant might be asked to perform a task and talk out loud to shed insights on which menu options were selected and why (Nielsen et al., 2002). These studies are conducted to make decisions about different choices during the design. Formative usability studies can be conducted on various pieces of the software and at different stages of development. The earlier a problem can be identified and solved, the more time and re-coding effort can be saved.

As an example, a study of an early personal digital assistant was conducted by user-experience personnel at Intel (Scholtz et al., 1997). The device itself had not yet been developed, so the group used a Personal Computer Memory Card International Association (PCMCIA) card that was designed as an external memory card for the laptop computer. The actual size of the personal digital assistant was to be the same as this device so that it could easily be placed into a laptop to be updated. The device would have a touch screen so that the user could remove it to take on a trip and access various contact information via the touch screen. The test involved, among other things, testing whether users wanted to hold it in portrait or landscape mode and observing how users interacted with interaction controls placed in different locations. Because the device did not exist, the different designs were printed on paper and secured to the PCMCIA cards with tape.

Summative usability evaluations are conducted to document a product's usability before delivery. A larger number of participants are used (between 10 and 20) with all participants assigned the same tasks. Observers watch the participants and log the amount of time it takes for each task and whether the task can be completed without or with help. After the task is completed, the user is asked for her subjective rating of the software. Typical metrics collected are efficiency, effectiveness, and user satisfaction. Efficiency is the time it takes the user to do a particular task. Effectiveness is the percentage of users who can complete the task unaided and the percentage of those who can complete the task with some help. User satisfaction is measured through a standard user experience questionnaire to determine how well users liked the experience with the software (UEQ-Online, ND; Brooke, 1996; Chin et al., 1988).

An International Organization for Standardization (ISO) details what needs to be reported during a summative usability evaluation. This standard allows companies interested in purchasing software to request this information from the company delivering the software (ISO, 2006). This standard also documents the Common Industry Format developed at the National Institute of Standards and Usability (Scholtz, 2000).

5.1.4   A/B STUDIES

Another type of evaluation is a comparison of two systems using the same tasks and same data to determine the better system according to a number of metrics. In the commercial world, this evaluation might compare the current version of the software with the newly developed software. Again, these studies involve users using the software on specified tasks. Tasks used in this type of testing may focus on new tasks or tasks that have been considerably improved in the new version. Companies may even consider using the results from this type of test to entice customers to upgrade to the latest version. Metrics used in this type of study are usually efficiency, effectiveness, and user satisfaction.

5.1.5   FIELD STUDIES

As most usability studies are done in laboratory spaces, we also need to look at how the software products fare in more realistic work environments. A more realistic environment provides insights into how various software products are used, including the length of time spent and different sequences of products used. Field studies are often carried out on beta releases or actual releases to determine how the software functions with real data and works in conjunction with other software products in the users' environments. Again, these studies rely on having users use the software, but in this case with their own data and executing their own tasks. While it is not feasible for researchers to spend weeks or even days observing users in the field, a number of techniques can be employed to gather the pertinent information. Often it is feasible to log users' interactions with the product so that researchers can study the logs. Diary-type studies ask users to record in some fashion their use with the product, noting both successes and failures. Researchers can then obtain these reports, conduct some detailed follow-ups with end users, and compile this information, noting problems that need to be addressed in subsequent releases or future versions of the software (Dix, 2009; Shneiderman and Plaisant, 2006).

CHAPTER 6

Evaluation Needs for Visual Analytics

Visual analytics present a more complex evaluation challenge than typical usability assessments for software applications. As seen in the nested blocks model, not only are there various parts to evaluate in visual analytics but also there are dependencies between these blocks. That is, changes made as a result of evaluation results in one block can affect the design and implementation of other blocks. Before discussing evaluation techniques for visual analytics, we will first review some of the recommendations made in the original visual analytics agenda (Thomas and Cook, 2005). The book recommends that research on evaluation methodologies should be a part of visual analytics research and development agenda to help focus the research efforts and facilitate the transition of the technology. Thomas and Cook cite the following benefits of evaluation.

- Verify research hypotheses.

- Encourage research in a particular area.

- Increase communication among academia, industry, and government.

- Compare technical approaches.

- Determine if program goals being achieved.

They also provide the following three general recommendations for evaluation.

1. All visual analytics research efforts must address the evaluation of their research results.

2. An infrastructure should be created to support evaluation of visual analytics research and tools.

3. A knowledge base characterizing visual analytics capabilities based on the results of evaluations should be developed.

Today, the efforts in the Visual Analytics Science and Technology (VAST) Challenge have been reasonably successful in addressing the infrastructure challenge (Cook et al., 2014). We will discuss how VAST Challenge data and submissions can be used for self-evaluation in a later section. There is still much work to do to address the other two recommendations. In the following sections, we discuss the efforts to modify existing HCI evaluation techniques where needed to address the needs of visual analytics evaluation. We first consider several additional recommendations from Thomas and Cook (2005).

> **Recommendation 3.1:** Conduct research to formally define the design spaces that capture different classes of visualizations. (p. 72)

> **Recommendation 3.2:** Develop a set of scientifically based cognitive, perceptual, and graphic design principles for mapping information to visual representations. (p. 73)

> **Recommendation 3.3:** Create a new science of interaction to support visual analytics. The grand challenge of interaction is to develop a taxonomy to describe the design space of interaction techniques that supports the science of analytic reasoning. We must characterize this design space and identify under-explored areas that are relevant to visual analytics. Then, R&D should be focused on expanding the repertoire of interaction techniques that can fill those gaps in the design space. (p. 76)

Recall that the goal of visual analytics is to support the analytic reasoning methodology, a process referred to as "sensemaking." Sensemaking has different tasks that need to be supported and different interactions with the data to support the sensemaking task, with attention paid to the cognitive and perceptual issues in designing the visualizations and accompanying interactions. Therefore, the user interface needs to accommodate this process, which might involve the ability to undo steps in the process if the analyst approaches a dead-end along a specific path and the ability to save information for incorporation into the final analytic report.

Currently, conducting user studies of the different types previously described and evaluations of visual analytics systems and/or components is problematic and consequently does not commonly occur. In an informal review of 15 conference papers presented at the VAST session of Vis 2016, only 2 papers reported conducting A/B. In these A/B studies, researchers most often compared a visual analytics tool with a more traditional tool. Another paper reported interviewing recent users of the systems, while 3 other papers reported on feedback collected while demonstrating their systems to expert users. One paper reported using initial user input to develop guidelines for development. One technique that was more frequently used is developers stepping through a task that users would most likely do. While it is reassuring that visual analytics researchers and developers recognize the need to obtain user feedback, obtaining that feedback earlier and more frequently during research and development would produce more usable and useful products more efficiently. However, to make that happen, some of the typical HCI evaluation methodologies need to be modified to make them suitable to early evaluations. In addition, there are more areas in visual analytics that need to be evaluated. In the following sections, we discuss modifications that are necessary to the typical HCI methods of evaluation and how those should be incorporated in the evaluation of visual analytics systems. Evaluations for new components in visualizations also are discussed.

6.1 MODIFYING TRADITIONAL HCI EVALUATION TECHNIQUES FOR VISUAL ANALYTICS

While all of the traditional HCI evaluation techniques can be applied to visual analytics systems, many of them require some modification. Furthermore, there are additional components that need to be evaluated in visual analytics systems, specifically data representations, algorithms, and support for sensemaking. As visual analytics systems support analysis, this support should also be evaluated. In the case of other techniques, the change may be the standard against which we judge the system or the metrics used in the evaluation. In others, it may be the procedure for the study itself that is somewhat different or that the data collected to analyze the results may differ. As suggested earlier, we may need to revise our HCI interaction techniques to support evaluation of the analytic reasoning process. There is the belief in visual analytics that giving users the ability to interact with visual representations of their data will lead to more insights. Research in this area has been termed "the science of interaction" (Pike et al., 2009). This term encompasses both lower-level interaction controls with the user interface and the higher-level interaction of the user with the problem space. Within this science of interaction concept, designers of visualizations need to consider the understandability of visualizations and the interactions of visualizations within the design of the visualizations, instead of as an add-on. Given this guidance, we first discuss new components that need to be evaluated and what those techniques should encompass. In the following sections, we discuss the modifications to typical HCI evaluation techniques to support evaluation of visual analytics systems.

6.2 NEW COMPONENTS

Data is an integral part of any visual analytics system. It is essential that the representation of any data ultimately support the analyst in her sensemaking activities. The nested blocks model (Meyer et al., 2015) contains blocks for both techniques and algorithms. The technique block includes the tasks of data encoding and interactions with the data. This block is designed and needs both guidelines for design and user studies to evaluate. The algorithm block deals with the task of rendering the data and any changes to the data on the screen. Again, both guidelines and user studies are essential in evaluating these blocks.

6.2.1 TECHNIQUES: DATA ENCODING AND INTERACTIONS

Data is at the heart of visual analytics and can take many forms, including text, numeric, images, and videos. Data may be homogeneous or heterogeneous. It may be clean or dirty. There are several considerations with respect to data: how much time and labor is needed to import data into the system; does the encoding provide the appropriate metadata to the analyst to support sensemaking; and will the process for encoding scale to the amount of data expected? Interactions with the data also need to be evaluated. Can the change brought about by the interaction be sufficiently displayed for the user to understand the result of this interaction?

Micallef et al. (2012) provide an excellent discussion of a user study done to look at different representations of Bayesian reasoning. They looked at six different visualizations for probabilistic reasoning and were unable to confirm previous guidelines. They did find that providing simple visualizations for textual explanations increased the accuracy of reasoning in their user studies. Today, while guidelines are used in designing the data encoding and interactions, few user studies are performed. As more novel visualizations are being created and visualizations are being used in more domains, it is essential that more user studies be done to ensure that the population the visualizations are being created for can effectively use them.

6.2.2   ALGORITHMS

The task in the algorithm block is to render on the screen the data and interactions from the techniques block. For example, the time it takes to render data may affect choices of data encoding and interaction. As data increases and algorithms increase in complexity, the time it takes to render the results of an interaction may suffer and may well affect the analyst's work. Stolper et al. (2014) discuss different approaches to this problem, including "progressive analytics" in which a partial analysis is returned to the user. However, it is essential to ensure that the design of the techniques block is such that this makes sense to do and that the user is able to understand the partial analysis enough to decide if this path is one she wants to pursue.

Stolper et al. (2014) provide guidelines for progressive analysis divided into those that apply to the analytics components and those that apply to the visualizations. Stopler (2014) found that analytics components should provide increasingly meaningful partial results as the algorithm executes, users should be allowed to focus the algorithm into subspaces of interest, and users should be able to specify subspaces to ignore. Visualizations should be designed to minimize distractions, provide cues to where new results have been found by analytics, support on-demand refresh, and provide an interface to allow users to specify where the analytics should focus and what part of the problem space should be ignored.

Evaluation of algorithms in visual analytics is also necessary to show that interactions with those visualizations produce results that users can interpret successfully. The cognitive and perceptual loads on users to determine what happens when they interact with the data should allow them to quickly interpret the outcome. As visual analytics move to support extremely large-scale data, the issue of the amount of data that can be shown, both with respect to the display space and also the time to render the data needs to be evaluated.

6.2.3   SENSEMAKING

Sensemaking refers to the highly iterative process that analysts go through to locate information, construct the appropriate representations, and eventually make sense of the information. During this process, analysts must have the ability to try out new representations, delete them when they are deemed inappropriate, and resurrect them when it becomes apparent that they might work after all. During sensemaking, an analyst may come to a dead-end after trying one possible alternative. At this point, can the software support the analyst in backing up and trying another approach? Currently, many visual analytic systems only provide access to the data through interactions and visualizations. Therefore, analysts must use other software to keep track of their information collected, sensemaking representations created, and the process they are using. Some visual analytics software is beginning to provide spaces for analysts to capture information and notes about what they have done and plan to do. The examples provided in the earlier section on Examples of Visual Analytics Tools show support for sensemaking as well as support for presentation of the analysis.

If sensemaking is to be supported in a visual analytics system, then the following requirements should be applied.

- The space allowed for analysts for sensemaking should provide an easily configured system as analysts will use notations that work for them.

- Analysts will iterate over the different representations they create so saving representations that can easily be restored is essential.

- Ideally, an evaluation of a well-created sensemaking space would show that an analyst using it would spend far less time in another application (or even using paper) and would be able to quickly re-create previously used representations and back up to take another path.

- The ways to accomplish these tasks should be obvious in the sensemaking space.

- Sensemaking evaluations can be done in early studies as well as in field tests. Early tests can be accomplished in laboratory to have the targeted users look at designs and interactions possible in the sensemaking space to determine if these would be helpful.

- Field studies can provide many more insights into how useful sensemaking spaces are for a variety of analysts, but as those are late in the development cycle, there is a significant cost for correcting or adding features.

A bright spot is that as researchers work with users in different domains, a body of knowledge will be accumulated that provides some initial rationale for designs for sensemaking. Pirolli and Russell (2011) in a special issue of "Human-Computer Interaction" provide several papers that discuss sensemaking in a number of disparate domains, including the control room of a race car venue (Wahlström et al., 2011), a team-based legal investigation (Attfield and Blandford, 2011), collaborative web searching (Paul and Morris, 2011), and educative training (Butcher and Sumner, 2011). Kang and Stasko (2012) discuss students in intelligence analysis tasks.

6.3   MODIFICATIONS TO HCI EVALUATION TECHNIQUES

In this section, we discuss how to apply guidelines, heuristics, A/B studies, and field studies to visual analytics systems and any modifications that may be required for these techniques.

6.3.1   GUIDELINES

As we noted earlier, many different categories of guidelines need to be used in designing and evaluating visual analytics systems, including visualizations, interactions, and sensemaking. While guidelines for visualizations have existed in the information visualization field for some time, these need to be augmented with guidelines for interaction with visualizations. Guidelines for enabling sensemaking are more about keeping user interface distractions from the user to enable the analyst to focus on reasoning about analytics.

Guidelines for Visualizations

A number of guidelines for visualizations are found in (Ware, 2012; Heer and Agrawala, 2008; Elmquist and Fekete, 2010). Ware (2012) provides a wide range of topics pertaining to visualizations, including color, visual attention, static and moving pattern, visual objects and data objects, space perception, and interacting with visualizations. Elmqvist and Fekete (2010) provide six guidelines for designing aggregated visualizations.

1. **Entity budget.** Maintain an entity budget: The data displayed on the screen should be controlled to avoid overloading the user.

2. **Visual summary.** Aggregates should convey information about underlying data: The user should be presented with an overview of the data but should not be overloaded with the entire set of data.

3. **Visual simplicity.** Aggregates should be clean and simple: Reduce clutter and simplify the overview.

4. **Discriminability.** Aggregates should be distinguishable from data items: Introduce novel visualizations for the data aggregates to differentiate them from the data items.

5. **Fidelity.** Beware that abstractions may lie: Make it clear to the user if data aggregates are used and thus convey information about the data that is not true (such as size of the data).

6. **Interpretability.** Aggregate items only so much so that the aggregation is still correctly interpretable within the visual mapping: This is related to Fidelity. Ensure that data is aggregated to help the user interpret the data, not just to reduce the size.

Ellis and Dix (2007) provide a taxonomy of techniques to reduce clutter in information visualizations. This taxonomy includes techniques to deal with appearance, spatial distortion, and time. For appearance, the techniques include sampling, filtering, change point size, change opacity, and clustering. For spatial distortion, the techniques include point/line displacement, topological distortion, space-filling, pixel-plotting, and dimensional ordering. For time, the technique is animation. All of these techniques are supported with numerous research efforts.

However, as we already pointed in the HCI Evaluation Techniques section, the use of guidelines as an evaluation technique is extremely difficult. Adding more guidelines to accommodate the various visualizations is even more difficult. Heer et al. (2010) provide a number of different visualizations along with the type of data and tasks that the visualizations are particularly suited to display.

Guidelines for Interactions

While a number of interaction techniques currently exist for information visualization, they exist at a relatively low level, i.e., filter, zoom, drill down, rotate, linking, animating (Shneiderman, 1996; Chuah and Roth, 1996; Dix and Ellis, 1998; Keim, 2002); Yi et al. (2007) argue that these low-level interaction techniques are not appropriate to use in developing a taxonomy of interaction techniques to support analytic reasoning and propose a set of higher-level tasks described below.

- **Select:** Mark something as interesting.

- **Explore:** Show me something else.

- **Reconfigure:** Show me a different arrangement.

- **Encode:** Show me a different representation.

- **Abstract/Elaborate:** Show me more or less detail.

- **Filter:** Show me something conditionally.

- **Connect:** Show me related items.

Ware (2012) discusses three categories of interaction: visual-manual control, view refinement and navigation, and the problem-solving loop. Ware notes that visualizations for the human act as an external aid to cognition and can facilitate the human in their problem solving. He notes that three levels of interaction need to be considered when designing and implementing interactive visualizations. The first level is interacting with various data elements. At this level, reaction time, positioning and selection, and detecting infrequently appearing targets are the interactive tasks. Guidelines concerning reaction times, consistency of controls, and support for vigilance are of primary importance.

The second level is the view refinement and navigation loop. Here the human needs to navigate through the model of the data and possibly refine her view. Navigation metaphors apply here, such as walking, flying, the world-in-hand, and eyeball-in-hand. Each of these metaphors needs to provide the proper affordances to users. There are many considerations here, such as the temporal scale needed, distortion techniques to provide room for data points of specific interest, zooming techniques, and multiple windows. All of these techniques have associated guidelines.

The third and highest level is the problem-solving loop. Visualizations are an external memory extension that should facilitate users in their problem-solving or sensemaking efforts as discussed below.

Guidelines for Sensemaking

Ware (2012) discusses the human memory system, iconic memory, short-term memory, and long-term memory and how visualization systems can help overcome issues in human memory. Besides having an external representation to augment human memory, the ability to create a sketch of our reasoning, such as a relationship diagram based on our findings, would be an extremely useful function for a visual analytics system. Green et al. (2008) present guidelines that address the cognitive activities of the user and note that the main concern in this regard is to keep the manipulation and interactions in the system such that the user can maintain focus on analysis, not the user interface. Specific guidelines are as follows.

- Provide multiple organizational views of the same information. The interaction should be balanced to occur automatically in all views.

- The interaction should be directly accomplished. The user should not have to pull down a menu or take other intermediate actions to accomplish an interaction.

- Facilitate the flow of human reasoning. Temporal pace of mental activity should optimize cognitive resources and present information in a relevant context.

- Intimate interaction. On-screen tools should not require additional cognitive focus but should be transparent and supportive of holistic cognition.

An example is an interface where you could select a word or phrase and use that to search by example. The queried information becomes a reasoning artifact. The above guidelines should help to ensure that the visual analytics system keeps the user in the cognitive zone as much as possible.

Green et al. (2009) provide a framework that extends current perceptual models of human processing. Based on this framework, they propose guidelines to extend the human-computer capabilities for reasoning in visual analytics systems. They propose guidelines for discovery and knowledge building, search by example and pattern, and creation and analysis of hypotheses.

Patterson et al. (2014) note that "there is a set of leverage points a given visualization designer might exploit in order to influence human cognition in the visual analytics process." Six leverage points from their study are capture exogenous attention, guide endogenous attention, facilitate chunking, aid reasoning with memory models, aid analogical reasoning, and encourage implicit learning. For each of these leverage points, the authors provide discussions and examples of techniques to achieve them. These leverage points could also be turned into guidelines or heuristics to evaluate the cognitive load of a visual analytics system.

In the Analytics section, we discussed different methods that analysts use in conducting analysis. Support for these different types of methods is also important, such as support for Analysis of Competing Hypotheses (Heuer, 1999; Richards and Pherson, 2010). The ACH Software Tool (PARC, The ACH Software Tool, 2017) developed by the Palo Alto Research Center, is an example of a tool developed to support that that particular technique by allowing the analyst to provide her hypotheses and then provide evidence that either supports or refutes a particular hypotheses. The analyst can set up a specific system to weight the evidence, thereby providing an automatic assessment of which hypotheses are more likely to be true given the discovered evidence.

Kang et al. (2009) conduct a number of studies with students using their Jigsaw system. They find that the analysts use four strategies: Overview, Filter, and Detail (OFD), Build from Detail (BFD), Hit the Keyword (HTK), and Find a Clue, Follow the Trail (FCFT). The researchers use their analysis of these strategies to suggest things that should be analyzed in a visual analytics system to facilitate the processes.

- Does the tool help to provide information scent to aid in finding initial clues?

- Does it guide the analyst to follow the right trail, without distraction?

- Does it support different strategies (sequences) for the sensemaking process?

- Does it allow flexibility for the analyst to organize information?

- Does it help to find appropriate next steps when the analyst encounters a dead-end in the sensemaking process?

- Does it facilitate further exploration?

These questions could certainly be reworded as guidelines for systems to support sensemaking. At present, researchers are focusing on building visual analytics systems and not evaluating them from the point of view of sensemaking. The guidelines here are certainly a good start to investigating different systems to determine techniques for supporting these activities and strategies.

We mentioned that even in typical HCI evaluations, specific guideline reviews are seldom done due to the sheer number of guidelines available. There are even more guidelines for visual analytics systems. Additionally, as more visualization techniques are developed, more guidelines will be developed. While it is essential to develop these guidelines for the designers, actual specific guideline evaluations are too complex to conduct. However, it will be essential to incorporate this information into heuristic evaluations as will be discussed in the section below.

6.3.2   HEURISTICS FOR VISUAL ANALYTICS

As in typical HCI evaluations, heuristics are an inexpensive way to evaluate software starting early in the design phase. However, the issue in visual analytics systems is that currently there is not an accepted set of heuristics to use. Forsell and Johansson (2010) provide a set of heuristics for information visualization using the same methodology as Molich and Nielsen (1990). They use several proposed sets of heuristics to arrive at the ten below that account for the majority of usability problems identified in information visualization.

1. **Information coding.** Perception of information is directly dependent on the mapping of data elements to visual objects. This should be enhanced by using realistic characteristics/techniques or the use of additional symbols.

2. **Minimal actions.** Concerns workload with respect to the number of actions necessary to accomplish a goal or a task.

3. **Flexibility.** Flexibility is reflected in the number of possible ways of achieving a given goal. It refers to the means available to customization in order to take into account working strategies, habits, and task requirements.

4. **Orientation and help.** Functions like support to control levels of details, redo/undo of actions, and representing additional information.

5. **Spatial organization.** Concerns users' orientation in the information space, the distribution of elements in the layout, precision and legibility, efficiency in space usage, and distortion of visual elements.

6. **Consistency.** Refers to the way design choices are maintained in similar contexts, and are different when applied to different contexts.

7. **Recognition rather than recall.** The user should not have to memorize a lot of information to carry out tasks.

8. **Prompting.** Refers to all means that help to know all alternatives when several actions are possible depending on the contexts.

9. **Remove the extraneous.** Concerns whether any extra information can be a distraction and take the eye away from seeing the data or making comparisons.

10. **Dataset reduction.** Concerns provided features for reducing a dataset, their efficiency, and ease of use.

These ten heuristics were assembled from experiments conducted using heuristics from Craft and Cairns (2005); Shneiderman (1996); Freitas et al. (2002); Nielsen (1994a); Nielsen Norman Group Website; Zuk and Carpendale (2006); Scapin and Bastien (1997); and Amar and Stasko (2004).

Väätäjä et al. (2016) used these heuristics and had several evaluators apply them to an actual system. Five evaluators applied these heuristics to a specified system. They were asked to provide a list of usability issues they found in the system and then rate each heuristic as to how applicable it was to each specific problem. After they finished, they were asked about the heuristic methodology and what if any additional heuristics might be needed. The evaluators found seventeen issues that were not addressed by any of the ten heuristics. The heuristics **Information coding** and **Orientation and help** accounted for the greatest number of issues. There were also some software problems, such as freezing and some latency issues, as well as some misuses of color. The evaluators found that using the heuristic methodology was somewhat difficult, due mainly to their lack of domain knowledge of the application.

Scholtz (2011) analyzed the reviews from the 2009 VAST Challenge to determine what reviewers felt were important aspects that were or were not included in the submitted visual analytic systems. In addition, she had three analysts review five video submissions which they rated and commented on individually. Comments included giving users control over the speed of animations, use color coding carefully and explain the codes, do not make visualizations too dense, ensure that size and distance used for representations are easy to interpret, do not require repetitive interactions by the analyst when exploring visualization, if possible display the process the analyst is using in the user interface, and make any automation transparent to the user.

While Nielsen's heuristics have become commonly accepted for HCI design, there is no commonly accepted set of heuristics for visualizations. The research needed here is to take one or more proposed sets of heuristics and determine how well they apply to visual analytics systems. As noted earlier, novel visualizations are constantly being developed, so research is needed to determine if the proposed heuristics can be applied to those or if new guidelines for new novel visualizations need to be applied to specific problems to determine if additional heuristics are needed or if current heuristics can cover additional visualizations and sensemaking activities.

A question is how early in the software development can a heuristic evaluation be used. For example, cognitive issues might be dependent on having interactions with the data or at least having a "before" and "after" to show changes in a visualization due to an interaction. There is also a question as to how knowledgeable heuristics evaluators need to be in the domain of the visual analytics system. Hearst et al. (2016) conducted studies to determine how effective heuristics were for visual analytics systems and found that asking users questions about the visualizations correlated well with results from the heuristic evaluations. Conducting both kinds of evaluations revealed further benefits especially when discrepancies were discovered.

6.3.3   USABILITY STUDIES

Recall the earlier discussion of formative and summative studies for typical HCI evaluations. Both types of evaluations are also appropriate for visual analytics. For formative studies, the issue is to ensure that the study is realistic. That is, the task, data, and time frame used in the study should match what the user will actually do. Remember that in formative studies, we may be only testing functionality or a single module so it may be necessary to provide a story about how this will fit into the visual analytics system as a whole. The following are typical issues we might test in a formative visual analytics usability study.

- Can users find relevant documents or entities using our visualization?

- Given different interactions for a specific visualization, do the users know which ones to select to support a given task?

- Is support for saving results of an interaction acceptable to the user?

- Is the correct metadata supported in the visualization?

- Does the visualization scale to support the amount of data that the user may have to deal with?

- Are the visualizations easy for the analyst to interpret?

Summative usability studies are difficult to conduct with visual analytics systems due to the time that users may spend with the system and the amount of training that may be required to use the system effectively. Visual analytics systems are often used daily on one or more tasks. Datasets and tasks need to be realistic, and finding users who can spend days or weeks doing a usability study is difficult and expensive. Moreover, accuracy is not usually a metric associated with analysis, so determining how to measure the outcome of summative usability tests needs to be determined.

While no real modifications are needed, the design of both formative and summative usability studies needs to be carefully done to make it match the end users' data and tasks. And while formative studies can be successfully implemented, great care should be taken to make all presented materials as realistic as possible.

As summative testing can be an extensive process, it may be better to conduct either A/B studies or field studies. A/B studies can highlight differences between a current software tool and a new tool but participants will need sufficient training in the new tool to perform the necessary tasks. Field testing will provide more information, but procedures for data collection will need to be formulated. These two techniques are discussed below. If summative testing is done, obtaining data, tasks, and users is a concern. The metrics for the studies also need to be established and discussed with the user community to make sure they reflect the users' concerns.

6.3.4   A/B STUDIES

As with summative usability testing, A/B studies involve finding the right data and the right users, formulating the appropriate metrics, and conducting the study for the appropriate amount of time. Because an A/B study is a comparison of two systems, there must be either two sets of comparable users, one set using each system (a "between user study design") or balancing the order that users use the systems (a "within user study design"). A between user study has two different groups of users. One group uses one system; the other group uses the second system. A within user study has all users using both systems in a balanced order. For example, if there are 20 subjects using systems A and B, then 10 subjects would randomly be assigned to use system A first and system B second. The other 10 subjects would use system B first, and system A second. Either type of evaluation poses problems. If two sets of users are needed they must be comparable in domain expertise and training with each system. Balancing the order is also difficult as it is impossible not to "learn" from using the first system if both systems use the same data. It is sometimes possible to find two sets of data and tasks that are comparable so that each system is used with a different set. But it is still necessary to ensure that the users are trained to a reasonable level of competence on both systems, which will add greatly to the time needed for the study.

6.3.5   FIELD STUDIES

As with typical HCI field studies, there are challenges here. On the positive side, field studies will clearly show how well the visual analytics system performs in actual work. In actuality, there are issues in finding one or more analysts who will spend some period of time using the system and provide feedback on the use of the system. It is often not feasible to have evaluators watch some of the time while the analysts use the system. It is also challenging to provide a way for analysts to provide feedback that does not consume large amounts of their time.

The biggest stumbling block may be the access to the analyst, data, and/or tasks that are being done. Often, analysts work in areas where their data and work are proprietary or somehow restricted. If this is not the case, then in-person-visits or online interviews are a good method to collect information supplemented by such things as diary entries of what was done, for how long, any problems, any successes, etc. If relatively easy metrics can be obtained, either by logging from the software or by the analyst herself, some valuable information can be gleaned. For example, the time worked, the number of different visualizations used, and the time spent using each one could be possible counts to track. Again, there are not really any modifications needed for field studies but a number of challenges are involved in getting approval for these studies and getting users, systems, and data all set to go. If the visual analytics system becomes the main system for the analysts, then it is necessary to ensure that technical support can quickly respond should an issue arise. On the positive side, field studies yield a rich source of information; however, it may not be possible to provide solutions to all reported issues in a timely release.

6.4   USING THE NESTED BLOCKS MODEL FOR EVALUATION TRACKING

We introduced the nested blocks model and guidelines earlier. This model can be used to help understand the context in which the evaluations are done and the dependencies within the blocks and between the blocks.

The outermost level is the domain level that is defined by the users, including their particular domain, their purpose in using the visual analytics system, the measurements that are important to them, and the data they will be using. These blocks can be at many different granularities ranging from an entire domain to smaller subdomains that ask specific questions. In any case, this is the context that sets the design and evaluation of the lower-level blocks.

The abstraction level consists of blocks for data and blocks for tasks. While data blocks are designed, tasks are extracted from the domain block. The blocks that are designed are the ones that need to be evaluated.

The technique blocks are also designed. These blocks focus on the designs for representing the data visually and the design of the interactions with the data. These will both need to be evaluated but within the context of accomplishing the tasks in the abstraction level.

The algorithm level blocks are designed as well. These blocks focus on how to render the visualizations of the data in the technique block. In order to test these algorithms it is necessary to ensure that the amount of data being used in the evaluation is comparable to what is expected in actual use.

Guidelines exist both between blocks and within blocks. Both software development and evaluation need to consider both the context in which blocks are being developed or evaluated (between blocks) and how designs in different blocks at the same level affect each other, e.g., , data representations and interaction designs. The evaluations should be planned with the goal of tackling inner blocks first to minimize redesigns done on outer blocks that will result in changes and hence, another evaluation, to inner blocks. Meyer et al. (2015) provide excellent examples of how to evaluate and track user studies using a nested blocks diagram.

6.5   METRICS

There are two reasons for conducting software evaluations. The first is to find errors in the software that would cause an issue for the user. The second is to provide measures of the utility of the software to help users understand the benefit of that software to their tasks (Keim et al., 2008b). Visual analytics systems have many components: algorithms for encoding and manipulating the data, visualizations, interactions on the visualizations, and possibly support for sensemaking. We need to devise metrics for each component as well as the system as a whole.

The usual HCI metrics for usability are efficiency, effectiveness, and user satisfaction (Bevan, 1999). Efficiency is measured by the time it takes users to complete various tasks they are asked to do. Effectiveness is computed by noting the tasks that users complete by themselves without errors. Often, the number of tasks they are able to complete by using any help available in the tool is also counted. User satisfaction measures how well the users like the software and is based on their ratings to standardized questionnaires (Chin et al., 1988; Lewis, 1995; Lund, 2001). Depending on the system, learnability may also be a measure. Systems such as kiosks should require no learning at all and are what we call "walk up and use." Office systems often require some amount of training; therefore, the usability testing should be accomplished after users have completed the recommended training.

While these measures are useful for individual tasks, such as "filter the visualization to show only documents originating in country X," they are not relevant for overall measures of visual analytics systems for several reasons. Analysts working with visual analytics systems have deadlines to meet. These deadlines often determine when the analyst is "done" with her task. Effectiveness is also suitable for low-level tasks, but a higher-level task such as "provide a recommendation for investments in portfolio Y" is much more difficult to measure. Accuracy has also been used to measure how well a task is done. In many domains, there is no notion of accuracy if accuracy is defined as the results obtained by the analyst. Rather, the analysts are either providing a picture of a current status of X or providing a recommendation concerning what might happen in the future, given what is known now and what is expected in the future.

6.6   METRICS FOR VISUAL ANALYTICS

Visual analytics systems have used a measure of utility; how useful do users find this system for the work they need to do? In devising these metrics, it is important to understand the users' tasks and the assistance that would be useful. For example, if a user wants to read as many relevant documents as possible, then a system that helps her select the relevant documents from a given source would be very useful. Thus, a metric such as number of relevant documents found could compare the new software to what the user currently has available.

Kang et al. (2009) provide the following metrics from a study they conducted:

- the number of important documents viewed, relative to the entire collection;

- when the analyst first started creating representations such as notes and drawings;

- the quantity of representations created;

- amount of time and effort in organizing; and

- amount of time the analyst spent reading/processing essential information.

Saket et al. (2016) look at studies evaluating the user experience while using visual analytics systems to determine possibilities for metrics. They examine three user experiences: memorability, enjoyment, and engagement but note that there may be other user experiences that should also be considered. For each of these three user experiences, the authors provide examples of studies and note what is measured for each.

Memorability is the easiest to measure. Can users recall visualizations they have seen and what types of designs help this recall? Studies have found that visualizations that are memorable at a glance have memorable content. Text and titles also help with memorability. Pictorial images and redundancy can also assist with recall (Borkin et al., 2016).

Engagement is more difficult. Mahyar et al. (2015) propose a taxonomy that can be used to measure the different stages of engagement: expose, involve, analyze, synthesize, and decide. Studies in this area to date have measured such things as the number of interactions the user has with the data and the time spent on the visualization but have not addressed a deeper level of engagement. Today, no guidelines for designs to enhance user engagement exist.

Enjoyment is even more elusive. Csikszentmihalyi (1990) identifies six factors that encompass enjoyment: challenge, focus, clarity, feedback, control, and immersion. The few studies looking at these factors ask subjects to self-rate their enjoyment. The studies find that embellished visualizations tend to produce more enjoyment (Bateman et al., 2010; Li and Moacdieh, 2014).

Insight has also been suggested as a metric for visual analytics systems. North (2006) provides the following characteristics of insight.

- **Complex.** Insight is complex, involving all or large amounts of the given data in a synergistic way, not simply individual data values.

- **Deep.** Insight builds up over time, accumulating and building on itself to create depth. Insight often generates further questions and, hence, further insight.

- **Qualitative.** Insight is not exact, can be uncertain and subjective, and can have multiple levels of resolution.

- **Unexpected.** Insight is often unpredictable, serendipitous, and creative.

- **Relevant.** Insight is deeply embedded in the data domain, connecting the data to existing domain knowledge and giving it relevant meaning. It goes beyond dry data analysis, to relevant domain impact.

There are a number of issues with evaluating insight. First, insights change over time. An experienced analyst in a specific domain will not have the same insights as a novice analyst in the same domain. Typical comparison (A/B) studies are not suitable for measuring insights unless analysts possess the same experience using the various software packages, necessitating long training times for the experiment. While benchmark tasks might be considered, they too have drawbacks. All participants must follow the same set of directions and the questions asked must have relatively simple answers to allow comparisons. This approach does not allow participants the freedom to conduct more exploration that might lead to additional insights.

Morse et al. (2005) conducted an evaluation for an intelligence community project, Novel Intelligence from Massive Data (NIMD). This project had five different areas of research: modeling analysts and analytic processes, prior and tacit knowledge, hypotheses generation and tracking, massive data, and human information interaction. For the human information interaction research work, the metrics devised in conjunction with analysts and researchers are shown in Figure 6.1.

Efficiency
- Time/search
- Time/document read

Effort
- # documents accessed
- # documents read
- Document growth rate
- Document growth type (cut/paste vs. typing)

Accuracy
- Evidence used in analysis
- Number of hypotheses considered
- Average system rank of documents viewed

Confidence
- User confidence ratings of findings

Answer/Report Quality
- Quality of report
- Ranking of report

Cognitive workload
- Cognitive workload ratings (NASA TLX; Hart and Staveland, 1988)

Figure 6.1: Metrics used for Novel Intelligence from Massive Data Project. Based on Morse et al. (2005).

Confidence was used as a metric. Although an analyst working with massive amounts of data will not be able to assess that she has read all the relevant papers, the more papers she reads that support her hypothesis, the more confident she will be in her recommendation. Accuracy was a metric used, but the definition of "accuracy" was the number of hypotheses the analyst considered and the evidence she used in her analysis. Again, there is really no way to know what the "accurate" recommendation is.

In this project, an evaluation framework was used to develop the metrics (Scholtz and Steves, 2004). In this framework, the goals of the system are considered first. Within each goal, each project supporting the research defines metrics that measure its contribution to the overall system goals. The evaluation framework comprises five levels: the system goals, the evaluation objectives, metrics, and two levels of metrics—conceptual and specific. Depending on the particular research project, the same conceptual metric may be used, but the conceptual measure and the specific implementation of that measure may differ for different projects.

Recall that earlier we talked about critical thinking and reasoning and discussed standards that intelligence analysts apply to their products. These are also possibilities for metrics that can be applied to summative studies of visual analytics software. Can users of visual analytics software produce higher quality products?

This discussion of metrics is intended to provide ideas, but the actual metrics used should be elicited from the intended users of the software. These are the metrics used at the highest block of the nested blocks and guidelines model and can be qualitative or quantitative. Metrics used in studies in the lower blocks need to contribute to these higher-level metrics. For example, if a metric important to users is "increased confidence in an analysis," then metrics at lower levels might measure the number of documents returned from a search that are ranked highly relevant or a significant increase in the number of documents that the analyst is able to view in a given time frame. The result would be compared to the previous system used by the analyst.

6.6.1   AN INFRASTRUCTURE TO SUPPORT EVALUATION NEEDS: THE VAST CHALLENGE

In the introduction to Evaluation for Visual Analytics section, we mention that one of three recommendations was to develop an infrastructure to aid in evaluation of visual analytics systems. The VAST Challenge was developed to meet this need. The VAST Challenge was started in 2005 to provide researchers developing visual analytics systems with datasets and the associated tasks. Unlike daily tasks faced by most analysts, the VAST Challenge provided ground truth for the posed questions. In the first two challenges, the datasets were heterogeneous, which proved to be an issue as many student research projects focused on one kind of data. Therefore, in 2007, the VAST Challenge was modified and consisted of several mini-challenges, each focusing on one type of data. However, there was also a grand challenge that necessitated answering all the mini-challenges first and then using the combined information to answer the grand challenge. Participants could enter one or more mini-challenges only or all mini-challenges and the grand challenge.

The objective of the VAST Challenge was to provide examples of both data and tasks that were similar to what actual analysts face. Visual analytics experts and analysts are asked to judge the submissions, which consist of descriptions of the process used to arrive at the answer and screen shots of the visualizations used. Participants are given the experts' feedback and also are given kudos for outstanding aspects of their systems, such as novel visualizations, good explanations, good support for sensemaking, etc. Participants agree to allow their submissions to be the posted on the Visual Analytics Benchmark Repository (Benchmark Repository). The data and questions are made available upon request as are the answers. This information has been used in visual analytics classes and projects.

Using this information is also a good way for researchers to evaluate a visual analytics system during development. It is easy to see which challenges use what type of data and the overall questions that are being asked and the tasks that must be done to answer these questions. As there is ground truth available, it is also easy to determine how well a given visual analytics system works in uncovering that ground truth. Cook et al. (2015) outline the process that researchers should follow in using VAST Challenge information to help evaluate their systems. The approach basically consists of looking for the type of data available for each VAST Challenge and determining if their system can handle that type of data. Researchers should then look at the various mini-challenges using that data to determine what tasks are required. If a researcher's system can handle one of more of those tasks, the researcher should try to analyze the data for one selected task and examine how it performed. Researchers can compare their results with the ground truth and with the processes and visualizations that were submitted for that mini-challenge. In particular, researchers should look at the awards that were given for accuracy and visualizations. While this is an informal evaluation technique, developers who use their own systems in this way often find areas where improvement is needed.

CHAPTER 7

Current Examples of Evaluation of Visual Analytics Systems

Publications of e valuation of Visual Analytics system are from both research and the real world. The VAST Challenge is a good example of providing feedback to visualization designers as both analysts and visualization researchers provide feedback. In this section we provide some information about the state of the art, both in our research communities and in practice.

7.1   EVALUATION IN RESEARCH COMMUNITIES

Lam et al. (2012) note that not only do visualizations have to be evaluated, but they also need to be evaluated in the context of the analysis process they are designed to support. Lam et al. (2012) provides seven different scenarios and provides techniques for evaluation in each of these. The scenarios for evaluating data analysis include: understanding environments and work practices, evaluating visual data analysis and reasoning, evaluating communication through visualization, and evaluating collaborative data analysis.

The scenarios for understanding visualizations are evaluating user performance, evaluating user experience, and evaluating visualization algorithms. These scenarios were derived from 850 papers in the literature. For each of these scenarios, the authors provide common evaluation goals, approaches, and evaluation questions. They note that evaluation can and should appear in all stages of software development: pre-design, design, prototype, deployment, and redesign. For example, if we were interesting in evaluating user performance, the most common metrics would be performance time and task accuracy. Two questions to answer concern the users' visual perception and cognitive processing with different types of data encoding and interactions and secondly, how different types of visualization and interactions impact the users' performance. Methods for measuring this include controlled experiments and field data collection. Often a controlled experiment is used to "benchmark" a system. Often this will be used to determine how much better a new version of a system is compared to the previous version.

Isenberg et al. (2013) coded 581 papers using Lam et al.'s scenarios to determine if evaluation had changed over time. Figure 7.1 shows the percentage of evaluations that included human participants. While the use of human participants in visual analytics evaluation is just reaching 50% it is rising. The use of human subjects in the evaluation of information visualization has fallen from earlier days but hovers between 80% and 90%. The question is why don't more evaluations include human participants? The most reasonable explanation is the difficult task of obtaining expert users for extended periods of time to participate. And, of course, when users are participating in these studies, their daily work is not being done.

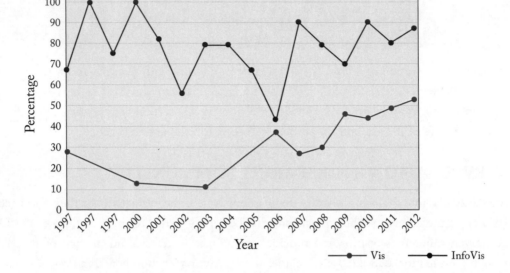

Figure 7.1: The percentage of evaluations that included human participants in the Vis Conference and the InfoVis Conference. Based on Isenberg et al. (2013).

Isenberg et al.(2013) looked more specifically at what types of evaluations and user data is being reported and they found that much of the early work with users of visualizations are not being reported in papers. That is, working with domain experts to understand their environments and work practices, assessing how visualization tools supports analysis and reasoning, evaluating communication through visualization, and evaluating collaborative data analysis. These activities and analysis tend to be more qualitative and are used primarily in the design of products. The conclusion section elaborates on how to encourage the reporting of these early observations.

7.2   INTEGRATING EVALUATION IN REAL-WORLD APPLICATIONS

Sedlmair et al. (2012) propose a methodology that combines the design of visualization software with the evaluation of how well it supports a real-world task. They outline nine steps used in the process: learn, winnow, cast, discover, design, implement, deploy, reflect, and write. Using this process actually sheds light on whether visualizations will be useful in the tasks that users will be doing with the software and if so, provides guidance on the design of those visualizations. There are several factors that need to be considered in this process. First, the task that users need to accomplish needs to be identified and characterized. How clear is this task? Will the task change over time or with different types of data? Secondly, where is the information that will be used in the task located? Is the information all located in the user's head or is it all digitized? Looking at these parameters helps to decide whether the task can be automated (clear task, all digital data) or whether a visualization is most likely not going to be useful(fuzzy task, little data outside of the user's head). The first steps (learn, winnow, and cast) consist of knowing this information, along with the different types of visualizations available, and deciding which of those are most probable to use. The different users need to be considered at this point as well. Assuming that a visualization system appears to be useful, then users must be consulted to discover what the actual requirements for the software are. Designs can then be considered and users can begin looking at various data abstractions and interactions. Prototypes can be developed and feedback from users gathered to help refine these designs. Formative testing can help refine the designs. Once the software is released to the users, feedback should still be collected to use in refinements of this software or to abstract into lessons learned for future software designs.

What are some of the problems that can happen in this process? Sedlmair et al. (2012) provide 32 "pitfalls." Some of the most severe are the lack of real data, the lack of a compelling reason for new software, ignoring the practices that real-world users currently have and like, not identifying the real end users, and not doing enough research up front to understand what visualizations are currently available. Sedlmair et al. (2011) conducted field studies in a large industrial company for over 3 years. Their task was to design and evaluate information visualization and visual analytics tools for the company. This involves expert users engaged in processes within the company. Thus the introduction of new tools has to fit into the processes easily, appeal to a variety of expert users, and offer a compelling benefit for adoption. There are a number of challenges and recommendations provided in the paper some of which are technical and some of which are political/organizational. Some of the biggest challenges are getting participants to work with, understanding their tasks and the variety of tasks the software needs to accommodate, obtaining real data to work with, and understanding the processes being used and the implications for inserting new software seamlessly into these processes. The collection of data in these studies and eventual publication are often complicated by confidentiality constraints. Sedlmair et al. (2011) described three phases of their work— pre-design, during design, and post-design—and noted the different types of HCI techniques used and the helpfulness of them. The most helpful techniques were contextual inquiries, focus groups (pre-design), design workshops, paper prototyping, user testing (design phase), long–term studies, and informal collaborations (post-design). They found early user observations to be much less helpful as it was difficult to understand what expert users were doing by observation only.

CHAPTER   8

Trends in Visual Analytics Research and Development

In the previous sections we discussed the work that needs to be done to development user-centered evaluations for current day visual analytics tools. However, visual analytics research is not standing still. As data increases and visual analytics methods appeal to more and more domains, user-centered evaluation needs to be aware of this work and ensure that evaluation methods are developed to help these new efforts to succeed.

8.1   COLLABORATIVE ANALYTICS

In many instances, analysis is done by teams and not just individuals. Often members take individual aspects of an issue to investigate. They then collaborate to combine their findings and integrate their recommendations.

Mahyer and Tory (2014) proposed the concept of Linked Common Work (LCW) and developed a prototype collaborative system in which members of a team were able to see externalizations of other team members' work during individual foraging and sensemaking. In a study contrasting their tool, CLIP, with a standard baseline, Mahyer and Tory found that this early awareness made the team sensemaking flow more smoothly. They used conversation analysis to produce both qualitative and quantitative metrics. Metrics used were tasks performance (relevant documents), communication, coordination, and awareness. They also noted what channels teams used for awareness (oral vs. CLIP) and the level of awareness at all time.

Scholtz and Endert (2014) provide a set of guidelines for collaborative systems. Their research showed that collaboration tools should: provide both content sharing and communication, support asynchronous use as well as synchronous use, support the integration of individual analyses as well as the discussion to produce a holistic product, should support both information gathering and analysis, help analysts keep in touch with other analysts, use subtle cues to notify analysts of additional information, provide support for analysts to use long-term standing queries to access information, and provide features for organizing evidence within a narrative.

8.2   STREAMING DATA

Crouser et al. (2017) discuss the electric power grid with its continual production of huge amounts of data that must be ingested by those users operating the grid in order to react and mitigate difficult situations. The authors discuss the challenges for visual analytics to deal with streaming data. Figure 8.1 shows the analysts' process while working with streaming data. As the data is continually changing, visualizations need to be designed to help ease the cognitive and perceptual load on the analyst. In dealing with the streaming data the users in the power grid spend much of their time monitoring various displays and trying to understand the current situation. The issue is that one display may change while the operator is focusing on understanding a different display, making the underlying concept completely different. The phases of the operator's process are orientation, reorientation, summary statistics, and scalability and approximation. The interactions and the resulting visualizations need to be displayed relatively quickly in order for the operator's interpretations to be made on current data.

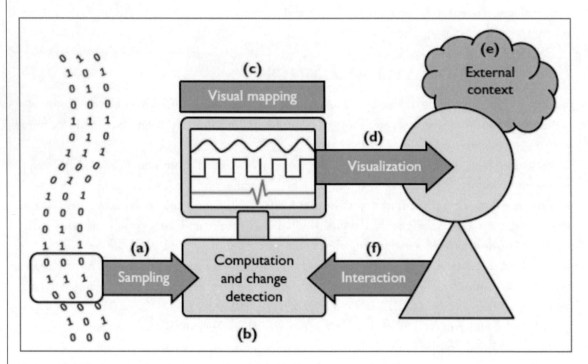

Figure 8.1: Streaming visual analytics lifecycle. (a) Data are sampled from various streams; (b) data are processed, relevant changes are detected; (c) the processed sample is mapped onto various visual dimensions; (d) the visualization is interpreted by the analyst; and (e) in the context of domain knowledge and other external information, which drives (f) interaction with the visualization and underlying model. Used with permission from Crouser et al. (2017).

- **Orientation:** The operator has access to the schedule of events for the day (any maintenance, substations being repaired, etc.), the weather forecast and the history of electric usage on a similar day. This is used to create a mental model of how the day might go, which is used in interpreting the streaming data. The issue is the speed at which this has to be done.

- **Reorientation:** The operator needs to continually monitor and re-adjust to developments that are happening. If major events occur, then the operator will need to readjust to those and quickly select and apply a mitigation strategy.

- **Summary statistics:** Each hour a new set of statistics are computed and displayed to the operator concerned the electrical needs so far and the projected electrical needs. These need to be responded to if the needs are significantly different than those projected earlier.

- **Scalability and approximation:** The refresh rate of the displays in an interactive system need to be at least 12 frames/second to be effective for user comprehension. The issue to select the subset of data that can be processed at this rate for the user to maintain situation awareness and react accordingly.

CHAPTER 9

Conclusions

In the above discussions of modifications to HCI evaluation techniques, we discussed how visual analytics system and component evaluations can be carried out today. However, if we think back to the Evaluation Needs for Visual Analytics Systems section and the three challenges posed, we can see that more research is needed. Thomas and Cook (2005) suggest that:

- all visual analytics research efforts must address the evaluation of their research results;

- an infrastructure should be created to support evaluation of visual analytics research and tools; and

- a knowledge base characterizing visual analytics capabilities based on the results of evaluations should be developed.

Freitas et al. (2014) advocated conducting user-centered evaluations for information visualization techniques and provided the following guidelines for accomplishing these evaluations:

- define the context of usage for evaluation prior to starting the evaluation;

- evaluation needs knowledge of the users and their goals and an understanding of which users are being supported with the software;

- evaluation needs to understand the tasks users do and the process they use to do this and which tasks are being supported in the software; and

- evaluation should be performed early in the design-development cycle.

These guidelines apply to the user-centered evaluations of visual analytics as well. Sharing the results of evaluation of different techniques should provide designers and developers of visualizations and visual analytics with more knowledge that can be incorporated in future systems.

Additionally, as in Isenberg et al. (2013) not all information is reported in information visualization and visualization papers. While studies that collect quantitative data are included in papers, the more informal and often qualitative studies that lead to design information are rarely included. This is likely due to reviewers' acceptance of statistically significant quantitative results and their skepticism of smaller studies with qualitative results. However, many of these studies and observations are useful to those in the field who do not have access to domain users. The results from these studies can help the community to generalize their thinking in design visual analytics systems. Thus, researchers should more carefully solicit comments from the users. This could be accomplished using more interviews or more in depth observations to ensure that the comments reflect the feeling of those in the user domain, not just a specific user or two.

The challenge for those of us doing the evaluation is to make sure our clients understand the importance of having their experts involved in the requirements, design, and implementation phases so that we can undertake evaluation efforts at each stage of the process. We need to ensure that we publish these evaluations so that the community can learn from our findings. We need to encourage researchers to conduct meta-analysis on these evaluations to help us understanding differences in domains and data.

We noted that the VAST Challenge is an ongoing effort creating an infrastructure to support evaluation of visual analytics research and tools. The repository provides researchers and practitioners with access to different scenarios, different visualizations, and different types of data. Awards that were given are listed so viewers may also have some insights into what features of the visual analytics system were viewed favorably by reviewers.

In addition, we should consider developing a knowledge base that designers, developers, and evaluation researchers can easily access and contribute to. Contributions of different methodologies used for evaluations including successes and failures would help advance the field of visual analytics. Furthermore, we need to encourage more research in evaluation, especially because of the evolving nature of visual analytics systems. Areas such as large displays, multiple screens, virtual and augmented reality, and collaborative efforts in analysis will need to be evaluated as well. The ever-expanding amount of data and heterogeneity of data will challenge visualizations and analysis as well.

The community needs more discussion of metrics for evaluating different types of visual analytics systems. There are several avenues to pursue for these discussions including BEliv (Beyond Time and Error) workshops held biannually and organizing discussion groups at the IEEE Vis Conference. The visual analytics community is moving ahead; user-centered evaluations need to keep pace to direct the research and development to those efforts that effectively support users. We need to encourage more user-centered evaluation work, including developing more heuristics for interactive visualization, developing techniques for evaluating components including interactions with different types of visualizations, and assessing cognitive challenges for users in understanding visualizations and the results of different interactions. We hope that this discussion will inspire more researchers to undertaken work in user-centered evaluations of visual analytics. Research in this area will help to transition more visual analytics systems into practice and to validate novel visualizations and interactions for new domains and tasks using visual analytics systems.

References

Albert, W. and Tullis, T. (2013). *Measuring the User Experience: Collecting, Analyzing, and Presenting Usability Metrics.* Newnes. 24

Amar, R. and Stasko, J. (2004). BEST PAPER: A knowledge task-based framework for design and evaluation of information visualizations. In *IEEE Symposium on Information Visualization, 2004. INFOVIS 2004.* IEEE. pp. 143–150. DOI: 10.1109/INFVIS.2004.10. 37

Apple Design Guidelines. https://developer.apple.com/design/ (accessed May 2017). 21

Attfield, S. and Blandford, A. (2011). Making sense of digital footprints in team-based legal investigations: The acquisition of focus. *Human–Computer Interaction*, 26(1-2), pp. 38–71. DOI: 10.1080/07370024.2011.556548. 32

Bateman, S., Mandryk, R. L., Gutwin,C., Genest, A., McDine, D., and Brooks. C. (2010). Useful junk? The effects of visual embellishment on comprehension and memorability of charts. *CHI '10 CHI Conference on Human Factors in Computing Systems Atlanta, GA*, April 10–15, ACM New York, NY. 42

Baumard, P. (1994). From noticing to making sense: using intelligence to develop strategy. *International Journal of Intelligence and Counter Intelligence*, 7(1), pp. 29–73. DOI: 10.1080/08850609408435236. 4

Bell, P. (1997). Using argument representations to make thinking visible for individuals and groups. In *Proceedings of the 2nd International Conference on Computer Support for Collaborative Learning* (pp. 10–19) International Society of the Learning Sciences. DOI: 10.3115/1599773.1599775. 9, 10

Benchmark Repository, http://hcil2.cs.umd.edu/newvarepository/ (accessed 2/10/2017). 45

Bevan, N. (1999). Quality in use: Meeting user needs for quality. *Journal of Systems and Software*, 49(1), pp. 89–96. DOI: 10.1016/S0164-1212(99)00070-9. 41

Borkin, M. A., Bylinskii, Z., Kim, N. W., Bainbridge, C. M., Yeh, C. S., Borkin, D., Pfister, H., and Oliva, A. (2016). Beyond memorability: Visualization recognition and recall. *IEEE Transactions on Visualization and and Computer Graphics*, 22(1). DOI: 10.1109/TVCG.2015.2467732. 42

Brooke, J. (1996). SUS-A quick and dirty usability scale. *Usability Evaluation in Industry*, 189(194), pp. 4–7. 25

Buchanan, R. (1992). Wicked problems in design thinking. *Design Issues*, 8(2), 5–21. The MIT Press. DOI: 10.2307/1511637. 7

Butcher, K. R. and Sumner, T. (2011). Self-directed learning and the sensemaking paradox. *Human–Computer Interaction*, 26(1-2), pp. 123–159. DOI: 10.1080/07370024.2011.556552. 32

Chin, J. P., Diehl, V. A., and Norman, K. L. (1988). Development of an instrument measuring user satisfaction of the human-computer interface. In *Proceedings of the SIGCHI Conference on Human Factors in Computing Systems,* ACM, pp. 213–218. DOI: 10.1145/57167.57203. 25, 41

Chuah, M. C. and Roth, S. F. (1996). On the semantics of interactive visualizations. In *Proceedings IEEE Symposium on Information Visualization'96,* IEEE, pp. 29–36. DOI: /10.1109/INFVIS.1996.559213. 33

CIA Library/Center for Intelligence. https://www.cia.gov/library/center-for-the-study-of-intelligence/ (accessed 03202017). 4

CIA Library/Intelligence History. https://www.cia.gov/library/publications/intelligence-history/oss/art02.htm (accessed 05132017). 3

CIA Library/Intelligence History. https://www.cia.gov/library/publications/resources/a-curators-pocket-history-of-the-cia/. Notes from our attic: A curator's Pocket History of the CIA, accessed March, 2017. 3

Cook, K., Grinstein, G., and Whiting, M. (2014). The VAST challenge: history, scope, and outcomes: An introduction to the specialissue. *Sage Journals*, 13(4), pp. 301–312. DOI: 10.1177/1473871613490678. 27

Cook, K. A., Scholtz, J., and Whiting., M. A. (2015). "A software developer's guide to informal evaluation of Visual Analytics environments using VAST Challenge information," *2015 IEEE Conference on Visual Analytics Science and Technology (VAST)*, Chicago, IL, pp. 193–194. DOI: 10.1109/VAST.2015.7347674. 45

Craft, B. and Cairns, P. (2005). Beyond guidelines: what can we learn from the visual information seeking mantra?. In *Information Visualisation, 2005. Proceedings. Ninth International Conference on*, IEEE, pp. 110–118. DOI: 10.1109/IV.2005.28. 37

Crouser, R. J., Franklin, L., and Cook, K. (2017). Rethinking visual analytics for streaming data applications. *IEEE Internet Computing*, 21(4), pp. 72–76. DOI: 10.1109/MIC.2017.2911428. 51, 52

Csikszentmihalyi, M. (1990). *Flow: The Psychology of Optimal Experience*. Harper Perennia, New York. 42

Dix, A. and Ellis, G. (1998). Starting simple: adding value to static visualisation through simple interaction. In *Proceedings of the Working Conference on Advanced Visual Interfaces* ACM, pp. 124–134. DOI: 10.1145/948496.948514. 3333Dix, A. (2009). *Human-computer Interaction* Springer U.S., pp. 1327–1331. 26

Dumas, J. S. and Redish, J. (1999). *A Practical Guide to Usability Testing.* Intellect Books. 24

Ellis, G. and Dix, A. (2007). A taxonomy of clutter reduction for information visualisation. *IEEE Transactions on Visualization and Computer Graphics*, 13(6), pp. 1216–1223. DOI: 10.1109/TVCG.2007.70535. 33

Elmqvist, N. and Fekete, J. D. (2010). Hierarchical aggregation for information visualization: Overview, techniques, and design guidelines. *IEEE Transactions on Visualization and Computer Graphics*, 16(3), pp. 439–454. DOI: 10.1109/TVCG.2009.84. 32

Fitts, P. M. (1954). The information capacity of the human motor system in controlling the amplitude of movement. *Journal of Experimental Psychology*, 47(6), p. 381. DOI: 10.1037/h0055392. 21

Forsell, C. and Johansson, J. (2010). An heuristic set for evaluation in information visualization. In *Proceedings of the International Conference on Advanced Visual Interfaces*, ACM, pp. 199–206. DOI: 10.1145/1842993.1843029. 24, 36

Freitas, C.M.D.S., Luzzardi, P.R.G., Cava, R.A., Winckler, M.A. A., Pimenta, S. M. and Nedel, L.P. (2002). Evaluating usability of information visualization techniques. In *Proceedings 5th Symposium on Human Factors in Computer Systems (IHC) 2002*, Brazilian Computer Society, pp. 40–51. 37

Freitas, C.M. D. S., Pimenta, M.S. and Scapin, D.L. (2014). User-centered evaluation of information visualization techniques: Making the HCI-InfoVis connection explicit. In W. Huang, editor, *Handbook of Human Centric Visualization*, Springer, pp. 315–336. DOI: 10.1007/978-1-4614-7485-2_12. 55

Green, T. M., Ribarsky, W., and Fisher, B. (2009). Building and applying a human cognition model for visual analytics. *Information Visualization*, 8(1), pp. 1–13. DOI: 10.1057/ivs.2008.28. 34

Green, T. M., Ribarsky, W., and Fisher, B. (2008). Visual analytics for complex concepts using a human cognition model. In *IEEE Symposium on Visual Analytics Science and Technology, 2008. VAST'08.* IEEE, pp. 91-98. DOI: 10.1109/VAST.2008.4677361. 34

Hart, S. G. and Staveland, L. E. (1988). Development of NASA-TLX (Task Load Index): Results of empirical and theoretical research. *Advances in Psychology*, 52, pp. 139–183. DOI: 10.1016/S0166-4115(08)62386-9. 43

Hearst, M. A., Laskowski, P., and Silva, L. (2016). Evaluating information visualization via the interplay of heuristic evaluation and question-based scoring. In *Proceedings of the 2016 CHI Conference on Human Factors in Computing Systems*, ACM, pp. 5028–5033. DOI: 10.1145/2858036.2858280. 38

Heer, J. and Agrawala, M. (2008). Design considerations for collaborative visual analytics. *Information Visualization*, 7(1), pp. 49–62. DOI: 10.1057/palgrave.ivs.9500167. 32

Heer, J., Bostock, M., and Ogievetsky, V. (2010). A tour through the visualization zoo. *Communications of the ACM*, 53(6), 59-67. DOI: 10.1145/1743546.1743567. 33

Heuer, R. J. (1999). *Psychology of Intelligence Analysis*. Lulu.com. 7, 35

HHS WebStandard for Order of Elements. http://webstandards.hhs.gov/guidelines/117 (accessed 11/09/2016). 22

HHS Webstandards and Guidelines. http://webstandards.hhs.gov/ (accessed 11/09/2016). 21, 22

Isenberg, T., Isenberg, P., Chen, J., Sedlmair, M., and Möller, T. (2013). A systematic review on the practice of evaluating visualization. *IEEE Transactions on Visualization and Computer Graphics*, 19(12), pp. 2818–2827. DOI: 10.1109/TVCG.2013.126. 47, 48, 55

ISO Standard, ISO/IEC 25062:2006 "Common Industry Format (CIF) for usability test reports"

Jones, M. D. (1998). *The Thinker's Toolkit: Fourteen Powerful Techniques for Problem Solving*. Crown Business. 7

Kang, Y. A. and Stasko, J. (2012). Examining the use of a visual analytics system for sensemaking tasks: Case studies with domain experts. *IEEE Transactions on Visualization and Computer Graphics*, 18(12), pp. 2869–2878. DOI: 10.1109/TVCG.2012.224. 32

Kang, Y. A., Gorg, C., and Stasko, J. (2009). Evaluating visual analytics systems for investigative analysis: Deriving design principles from a case study. In *IEEE Symposium on Visual Analytics Science and Technology, 2009. VAST 2009*. IEEE, pp. 139–146. DOI: 10.1109/VAST.2009.5333878. 35, 41

Keim, D., Andrienko, G., Fekete, J. D., Gorg, C., Kohlhammer, J., and Melançon, G. (2008a). Visual analytics: Definition, process, and challenges. *Lecture Notes in Computer Science*, 4950, pp. 154–176. DOI: 10.1007/978-3-540-70956-5_7. 12

Keim, D. A., Mansmann, F., Schneidewind, J., and Ziegler, H. (2006). Challenges in visual data analysis. In *Tenth International Conference on Information Visualization, 2006. IV 2006*. IEEE, pp. 9–16. DOI: 10.1109/IV.2006.31. 12

Keim, D., Mansmann, F., Schneidewind, J., Thomas, J., and Ziegler, H. (2008b). Visual analytics: Scope and challenges. *Visual Data Mining*, pp. 76–90. 41

Keim, D. A. (2002). Information visualization and visual data mining. *IEEE Transactions on Visualization and Computer Graphics*, 8(1), pp. 1–8. DOI: 10.1109/2945.981847. 33

Kent, S. (1955). The Need for an Intelligence Literature. Essay available at https://tinyurl.com/79n-3wku (accessed May,2017). 4

Klein, G., Phillips, J. K., Rall, E., and Peluso, D. A. (2007). A data/frame theory of sensemaking. In R. R. Hoffman (Ed.) *Expertise out of Context*. Psychology Press. 8

Lam, H., Bertini, E., Isenberg, P., Plaisant, C., and Carpendale, S. (2012). Empirical studies in information visualization: Seven scenarios. *IEEE Transactions on Visualization and Computer Graphics*, 18(9), pp. 1520–1536. DOI: 10.1109/TVCG.2011.279. 47

Lewis, J. R. (1995). IBM computer usability satisfaction questionnaires: psychometric evaluation and instructions for use. *International Journal of Human-Computer Interaction*, 7(1), pp. 57–78. DOI: 10.1080/10447319509526110. 41

Li, H. and Moacdieh, N. (2014). Is "chart junk" useful? An extended examination of visual embellishment. In *Proceedings of the Human Factors and Ergonomics Society Annual Meeting*, 58(1), SAGE Publications, pp. 1516–1520. DOI: 10.1177/1541931214581316. 42

Lindley, R. H. (1966). Recoding as a function of chunking and meaningfulness. *Psychonomic Science*, 6(8), pp. 393–394. DOI: 10.3758/BF03330953. 21

Lund, A. M. (2001). Measuring Usability with the USE Questionnaire12. *Usability Interface*, 8(2), pp. 3–6. 41

Mahyar, N., Kim, S. H., and Kwon, B. C. (2015). Towards a taxonomy for evaluating user engagement in information visualization. In *Workshop on Personal Visualization: Exploring Everyday Life*, (Vol. 3, p. 2). 42

Mahyar, N. and Tory, M. (2014). Supporting communication and coordination in collaborative sensemaking. *IEEE Transactions on Visualization and Computer Graphics*, 20(12), pp. 1633–1642. DOI: 10.1109/TVCG.2014.2346573. 51

Meyer, M., Sedlmair, M., Quinan, P. S., and Munzner, T. (2015). The nested blocks and guidelines model. *Information Visualization*, 14(3), pp. 234–249. 18, 20, 29, 40Micallef, L., Dragicevic, P., and Fekete, J. D. (2012). Assessing the effect of visualizations on bayesian reasoning through crowdsourcing. *IEEE Transactions on Visualization and Computer Graphics*, 18(12), pp. 2536–2545. 30

Microsoft Design Guidelines (Windows). MSND- Microsoft. https://msdn.microsoft.com/en-us/library/windows/desktop/dn688964(v=vs.85).aspx, (accessed October 2017). 21

Miller, G. (1956). The magical number seven, plus or minus two: some limits on our capacity for processing information. *Psychological Review*, 63(2), p. 81. DOI: 10.1037/h0043158. 21

Molich, R. and Nielsen, J. (1990). Improving a human-computer dialogue. *Communications of the ACM*, 33(3), pp. 338–348. DOI: 10.1145/77481.77486. 36

Moore, D. T. (2007). *Critical Thinking and Intelligence Analysis.* National Defense Intelligence College, Washington, D.C. DOI: 10.1037/e509522010-001. 4

Moore, D. T. (2011). *Sensemaking: A structure for an Intelligence Revolution.* National Defense Intelligence College, Washington, D.C. 4

Morse, E., Steves, M. P., and Scholtz, J. (2005). Metrics and methodologies for evaluating technologies for intelligence analysts. In *Proceedings Conference on Intelligence Analysis.* 43

Munzner, T. (2009). A nested model for visualization design and validation. *IEEE Transactions on Visualization and Computer Graphics*, 15(6). DOI: 10.1109/TVCG.2009.111. 18

NASA TLX, http://apps.usd.edu/coglab/schieber/psyc792/workload/Gawron-TLX.pdf (accessed September 2017). 43

Nielsen Norman Website, https://www.nngroup.com/articles/ten-usability-heuristics/ (accessed May 2017). 22, 37

Nielsen, J. and Molich, R. (1990). Heuristic evaluation of user interfaces. In *Proceedings of the SIGCHI Conference on Human Factors in Computing Systems*, ACM, pp. 249–256. DOI: 10.1145/97243.97281. 22

Nielsen, J. (1994a). Enhancing the explanatory power of usability heuristics. In P*Proceedings of the SIGCHI Conference on Human Factors in Computing Systems*, ACM. pp. 152–158. 22, 37

Nielsen, J. (1992). Finding usability problems through heuristic evaluation. In *Proceedings of the SIGCHI Conference on Human Factors in Computing Systems*, ACM. pp. 373–380. DOI: 10.1145/142750.142834. 24

Nielsen, J. (1994b). Heuristic evaluation. In *Usability Inspection Methods*. Nielsen, J. and Mack, R. L. (eds.), John Wiley & Sons:New York.

Nielsen, J., Clemmensen, T., and Yssing, C. (2002). Getting access to what goes on in people's heads?: reflections on the think-aloud technique. In *Proceedings of the Second Nordic Conference on Human-computer Interaction*, ACM, pp. 101–110. DOI: 10.1145/572020.572033. 24

North, C. (2006). Toward measuring visualization insight. *IEEE Computer Graphics and Applications*, 26(3), pp. 6–9. DOI: 10.1109/MCG.2006.70. 42

PARC. The ACH Software Tool. Palo Alto Research Center, a Xerox Company. http://www2.parc.com/istl/projects/ach/ach.html (accessed June 13, 2017). 35

Patterson, R. E., Blaha, L. M., Grinstein, G. G., Liggett, K. K., Kaveney, D. E., Sheldon, K. C., Havig, P. R., and Moore, J. A. (2014). A human cognition framework for information visualization. *Computers and Graphics*, 42, pp. 42–58. DOI: 10.1016/j.cag.2014.03.002. 35

Paul, R. and Elder, L. (2004). *The Miniature Guide to Critical Thinking: Concepts and Tools*. Foundation Critical Thinking. 4

Paul, S. A. and Morris, M. R. (2011). Sensemaking in collaborative web search. *Human–Computer Interaction*, 26(1–2), pp. 72–122. DOI: 10.1080/07370024.2011.559410. 32

Pezzotti, N., Lelieveldt, B. P., van der Maaten, L., Höllt, T., Eisemann, E., and Vilanova, A. (2017). Approximated and user steerable tsne for progressive visual analytics. *IEEE Transactions on Visualization and Computer Graphics*, 23(7), pp. 1739–1752. DOI: 10.1109/TVCG.2016.2570755.

Pirolli, P. and Russell, D. M. (2011). *Introduction to this Special Issue on Sensemaking*. Taylor and Francis. DOI: 10.1080/07370024.2011.556557. 31

Pike, W. A., Stasko, J., Chang, R., and O'connell, T. A. (2009). The science of interaction. *Information Visualization*, 8(4), pp. 263–274. DOI: 10.1057/ivs.2009.22. 29

Pirolli, P. and Card, S. (2005). The sensemaking process and leverage points for analyst technology as identified through cognitive task analysis. In *Proceedings of International Conference on Intelligence Analysis*, 5, pp. 2 4. 7, 9

Richards, H. J. and Pherson, R. H. (2010). *Structured Analytic Techniques for Intelligence Analysis*. Cq Press. 35

Russell, D. M., Stefik, M. J., Pirolli, P., and Card, S. K. (1993). The cost structure of sensemaking. In *Proceedings of the INTERACT'93 and CHI'93 Conference on Human Factors in Computing Systems*. ACM, pp. 269–276. DOI: 10.1145/169059.169209. 8

Saket, B., Endert, A., and Stasko, J. (2016). Beyond Usability and Performance: A Review of User Experience-focused Evaluations in Visualization. In *Proceedings of the Beyond Time and Errors on Novel Evaluation Methods for Visualization*, ACM, pp. 133–142. DOI: 10.1145/2993901.2993903. 42

Scapin, D. L. and Bastien, J. C. (1997). Ergonomic criteria for evaluating the ergonomic quality of interactive systems. *Behaviour and Information Technology*, 16(4-5), pp. 220–231. DOI: 10.1080/014492997119806. 37

Scholtz, J. (2011). Developing guidelines for assessing visual analytics environments. *Information Visualization*, 10(3), pp. 212–231. DOI: 10.1177/1473871611407399. 37

Scholtz, J. and Steves, M. P. (2004). A framework for real-world software system evaluations. In *Proceedings of the 2004 ACM Conference on Computer Supported Cooperative Work*, ACM, pp. 600–603. DOI: 10.1145/1031607.1031710. 44

Scholtz, J. C., Lockhart, P., Salvador, T., and Newbery, J. (1997). Design: no job too small. In *Proceedings of the ACM SIGCHI Conference on Human factors in Computing Systems*, ACM, pp. 447–454. DOI: 10.1145/258549.258837. 24

Scholtz, J. (2000). Common industry format for usability test reports. In *CHI'00 Extended Abstracts on Human Factors in Computing Systems*, ACM, pp. 301–301. DOI: 10.1145/633292.633470. 25

Scholtz, J. and Endert, A. (2014). User-centered design guidelines for collaborative software for intelligence analysis. In *2014 International Conference on Collaboration Technologies and Systems (CTS)*, IEEE, pp. 478–482. DOI: 10.1109/CTS.2014.6867610. 51

Sedlmair, M., Meyer, M., and Munzner, T. (2012). Design study methodology: Reflections from the trenches and the stacks. *IEEE Transactions on Visualization and Computer Graphics*, 18(12), pp. 2431–2440. DOI: 10.1109/TVCG.2012.213. 48, 49

Sedlmair, M., Isenberg, P., and Baur, D. (2011) "Information visualization evaluation in large companies: Challenges, experiences and recommendations." *Information Visualization* 10.3, pp. 248–266. DOI: 10.1177/1473871611413099. 49

Shneiderman, B. and Plaisant, C. (2006). Strategies for evaluating information visualization tools: multidimensional in-depth long-term case studies. *BEliv '06 Procedings of the 2006 AVI workshop on BEyond time and errors: on novel evaluation methods for information evaluation*, pp. 1–7. DOI: 10.1145/1168149.1168158. 26

Shneiderman, B. (1996). The eyes have it: A task by data type taxonomy for information visualizations. In *Proceedings, IEEE Symposium on Visual Languages,1996*. IEEE, pp. 336–343. DOI: 10.1109/VL.1996.545307. 33, 37

Shneiderman, B. (1987). *Designing the User Interface*, Addision-Wesley. 20

Smith, S. L. (1981). Design guidelines for the user-system interface of on-line computer systems: A survey report. In *Proceedings of the Human Factors Society Annual Meeting*, 25(1), SAGE Publications: Los Angeles, CA, pp. 509–512. DOI: 10.1177/1071181381025001131. 20

Smith, S. L. and Mosier, J. N. (1984). Design guidelines for user-system interface software (No. MTR-9420). Mitre Corp, Bedford, MA. 20

Smith, S. L. and Aucella, A. F. (1983). Design guidelines for the user interface to computer-based information systems (No. MTR-8857). Mitre Corp, Bedford, MA. 20

Stolper, C. D., Perer, A., and Gotz, D. (2014). Progressive visual analytics: User-driven visual exploration of in-progress analytics. *IEEE Transactions on Visualization and Computer Graphics*, 20(12), pp. 1653–1662. DOI: 10.1109/TVCG.2014.2346574. 30

Thomas, J. J. and Cook, K . A. (2005). *Illuminating the Path*, IEEE Computer Society Press. Los Alamitos, CA. 11, 27, 28, 55

UEQ-Online. "User Experience Questionnaire." http://www.ueq-online.org/ (accessed June 13, 2017). 25

Väätäjä, H., Varsaluoma, J., Heimonen, T., Tiitinen, K., Hakulinen, J., Turunen, M., Nieminen, H., and Ihantola, P. (2016). Information visualization heuristics in practical expert evaluation. In *Proceedings of the Beyond Time and Errors on Novel Evaluation Methods for Visualization*, ACM, pp. 36–43. DOI: 10.1145/2993901.2993918. 37

Wahlström, M., Salovaara, A., Salo, L., and Oulasvirta, A. (2011). Resolving safety-critical incidents in a rally control center. *Human–Computer Interaction*, 26(1-2), pp. 9–37. DOI: 10.1080/07370024.2011.556541. 32

Ware, C. (2012). *Information Visualization: Perception for Design*. Elsevier. 32, 33, 34

Wright, W., Schroh, D., Proulx, P., Skaburskis, A., and Cort, B. (2006). The Sandbox for analysis: concepts and methods. In *Proceedings of the SIGCHI Conference on Human Factors in Computing Systems*, ACM, pp. 801–810. DOI: 10.1145/1124772.1124890. 14, 15

Yi, J. S., ah Kang, Y., and Stasko, J. (2007). Toward a deeper understanding of the role of interaction in information visualization. *IEEE Transactions on Visualization and Computer Graphics*, 13(6), pp. 1224–1231. DOI: 10.1109/TVCG.2007.70515. 33

Zuk, T. and Carpendale S. (2006). Theoretical analysis of uncertainty visualizations. In *Proceedings of SPIE-IS&T Electronic Imaging*, SPIE, vol. 6060, 606007. 37

ADDITIONAL READINGS

Visual Analytics – General

Dill, J., Earnshaw, R., Kasik, D., Vince, J., and Wong, P. C. (Eds.). (2012). *Expanding the Frontiers of Visual Analytics and Visualization*. Springer Science and Business Media. DOI: 10.1007/978-1-4471-2804-5.

Ellis, G. and Mansmann, F. (2010). Mastering the information age solving problems with visual analytics. In *Eurographics*, Vol. 2, p. 5.

Keim, D., Andrienko, G., Fekete, J. D., Gorg, C., Kohlhammer, J., and Melançon, G. (2008c). Visual analytics: Definition, process, and challenges. *Lecture Notes in Computer Science*, 4950, pp. 154–176. DOI: 10.1007/978-3-540-70956-5_7.

Keim, D. A., Mansmann, F., Schneidewind, J., Thomas, J., and Ziegler, H. (2008d). Visual analytics: Scope and challenges. In *Visual Data Mining* Springer Berlin Heidelberg, pp. 76–90.

Kielman, J., Thomas, J., and May, R. (2009). Foundations and frontiers in visual analytics. *Information Visualization*, 8(4), pp. 239–246. DOI: 10.1057/ivs.2009.25.

Assessments and Evaluation of Visual Analytics

Adagha, O., Levy, R. M., and Carpendale, S. (2015). Towards a product design assessment of visual analytics in decision support applications: a systematic review. *Journal of Intelligent Manufacturing*, pp. 1–11.

Carpendale, S. (2008). Evaluating information visualizations. *Information Visualization*, pp. 19–45. DOI: 10.1007/978-3-540-70956-5_2.

Dasgupta, A., Lee, J. Y., Wilson, R., Lafrance, R. A., Cramer, N., Cook, K., and Payne, S. (2017). Familiarity Vs Trust: A Comparative Study of Domain Scientists' Trust in Visual Analytics and Conventional Analysis Methods. *IEEE Transactions on Visualization and Computer Graphics*, 23(1), pp. 271–280. DOI: 10.1109/TVCG.2016.2598544.

Kang, Y. A., Gorg, C., and Stasko, J. (2011). How can visual analytics assist investigative analysis? Design implications from an evaluation. *IEEE Transactions on Visualization and Computer Graphics*, 17(5), pp. 570–583. DOI: 10.1109/TVCG.2010.84.

Kang, Y. A., Gorg, C., and Stasko, J. (2009b). Evaluating visual analytics systems for investigative analysis: Deriving design principles from a case study. In *IEEE Symposium on Visual Analytics Science and Technology, 2009. VAST 2009.* IEEE, pp. 139–146.

Scholtz, J. (2006). Beyond usability: Evaluation aspects of visual analytic environments. In *IEEE Symposium on Visual Analytics Science and Technology, 2006*, IEEE. pp. 145–150. DOI: 10.1109/VAST.2006.261416.

Scholtz, J., Plaisant, C., Whiting, M., and Grinstein, G. (2014). Evaluation of visual analytics environments: The road to the Visual Analytics Science and Technology challenge evaluation methodology. *Information Visualization*, 13(4), pp. 326–335. DOI: 10.1177/1473871613490290.

Whiting, M. A., Haack, J., and Varley, C. (2008). Creating realistic, scenario-based synthetic data for test and evaluation of information analytics software. In *Proceedings of the 2008 Work-*

shop on Beyond Time and Errors: Novel Evaluation Methods for Information Visualization, ACM, p. 8. DOI: 10.1145/1377966.1377977.

Collaborative Visual Analytics Tools and Design

Andrews, C., Endert, A., and North, C. (2010). Space to think: large high-resolution displays for sensemaking. In *Proceedings of the SIGCHI Conference on Human Factors in Computing Systems,* ACM, pp. 55–64. DOI: 10.1145/1753326.1753336.

Bradel, L., Endert, A., Koch, K., Andrews, C., and North, C. (2013). Large high resolution displays for co-located collaborative sensemaking: Display usage and territoriality. *International Journal of Human-Computer Studies*, 71(11), pp. 1078–1088. DOI: 10.1016/j.ijhcs.2013.07.004.

Brennan, S. E., Mueller, K., Zelinsky, G., Ramakrishnan, I. V., Warren, D. S., and Kaufman, A. (2006, October). Toward a multi-analyst, collaborative framework for visual analytics. In *IEEE Symposium on Visual Analytics Science and Technology, 2006*, IEEE, pp. 129–136. DOI: 10.1109/VAST.2006.261439.

Heer, J. and Agrawala, M. (2008). Design considerations for collaborative visual analytics. *Information Visualization*, 7(1), pp. 49–62. DOI: 10.1057/palgrave.ivs.9500167.

Isenberg, P. and Fisher, D. (2009). Collaborative Brushing and Linking for Co-located Visual Analytics of Document Collections. In *Computer Graphics Forum*, 28(3), Blackwell Publishing Ltd., pp. 1031–1038. DOI: 10.1111/j.1467-8659.2009.01444.x.

Isenberg, P., Fisher, D., Morris, M. R., Inkpen, K., and Czerwinski, M. (2010). An exploratory study of co-located collaborative visual analytics around a tabletop display. In *IEEE Symposium on Visual Analytics Science and Technology (VAST), 2010*, IEEE, pp. 179–186. DOI: 10.1109/VAST.2010.5652880.

Morris, M. R., Lombardo, J., and Wigdor, D. (2010). WeSearch: supporting collaborative search and sensemaking on a tabletop display. In *Proceedings of the 2010 ACM Conference on Computer Supported Cooperative Work*, ACM, pp. 401–410. DOI: 10.1145/1718918.1718987.

Sarvghad, A., Tory, M., and Mahyar, N. (2017). Visualizing dimension coverage to support exploratory analysis. *IEEE Transactions on Visualization and Computer Graphics*, 23(1), pp. 21–30. DOI: 10.1109/TVCG.2016.2598466.

Sarvghad, A. and Tory, M. (2015). Exploiting analysis history to support collaborative data analysis. In *Proceedings of the 41st Graphics Interface Conference*, Canadian Information Processing Society, pp. 123–130.

Vogt, K., Bradel, L., Andrews, C., North, C., Endert, A., and Hutchings, D. (2011). Co-located collaborative sensemaking on a large high-resolution display with multiple input devices. *Human–Computer Interaction–INTERACT 2011*, pp. 589–604. DOI: 10.1007/978-3-642-23771-3_44.

Wallace, J. R., Scott, S. D., and MacGregor, C. G. (2013). Collaborative sensemaking on a digital tabletop and personal tablets: prioritization, comparisons, and tableaux. In *Proceedings of the SIGCHI Conference on Human Factors in Computing Systems*, ACM, pp. 3345–3354. DOI: 10.1145/2470654.2466458.

Sensemaking

Paul, S. A. and Reddy, M. C. (2010). Understanding together: sensemaking in collaborative information seeking. In *Proceedings of the 2010 ACM Conference on Computer Supported Cooperative Work*, ACM, pp. 321–330. DOI: 10.1145/1718918.1718976.

Russell, D. M., Stefik, M. J., Pirolli, P., and Card, S. K. (1993). The cost structure of sensemaking. In *Proceedings of the INTERACT'93 and CHI'93 Conference on Human Factors inComputing Systems*, ACM, pp. 269–276. DOI: 10.1145/169059.169209.

Weick, K. E. (1993). The collapse of sensemaking in organizations: The Mann Gulch disaster. *Administrative Science Quarterly*, pp. 628–652. DOI: 10.2307/2393339.

Weick, K. E. (1995). *Sensemaking in Organizations* (Vol. 3). Sage.

Nested Blocks Model

Munzner, T. (2014). *Visualization Analysis and Design*. CRC Press.

Rittel, H. W. and Webber, M. M. (1973). 2.3 planning problems are wicked. *Polity*, 4, pp. 155–169.

Interaction in Visual Analytics

Pike, W. A., Stasko, J., Chang, R., and O'connell, T. A. (2009). The science of interaction. *Information Visualization*, 8(4), pp. 263–274. DOI: 10.1057/ivs.2009.22.

Liu, Z. and Stasko, J. (2010). Mental models, visual reasoning and interaction in information visualization: A top-down perspective. *IEEE Transactions on Visualization and Computer Graphics*, 16(6), pp. 999–1008.

Yi, J. S., ah Kang, Y., and Stasko, J. (2007). Toward a deeper understanding of the role of interaction in information visualization. *IEEE Transactions on Visualization and Computer Graphics*, 13(6), pp. 1224–1231. DOI: 10.1109/TVCG.2007.70515.

Adoption of Visual Analytics

Chinchor, N., Cook, K., and Scholtz, J. (2012). Building adoption of visual analytics software. In *Expanding the Frontiers of Visual Analytics and Visualization,* Springer London. pp. 509–530. DOI: 10.1007/978-1-4471-2804-5_29.

Visual Analytics Applications

Kim, S. Y., Jang, Y., Mellema, A., Ebert, D. S., and Collinss, T. (2007). Visual analytics on mobile devices for emergency response. In *IEEE Symposium on Visual Analytics Science and Technology, 2007. VAST 2007,* pp. 35–42. DOI: 10.1109/VAST.2007.4388994.

Krishnan, M., Bohn, S., Cowley, W., Crow, V., and Nieplocha, J. (2007). Scalable visual analytics of massive textual datasets. In *Parallel and Distributed Processing Symposium, 2007. IPDPS 2007. IEEE International,* IEEE, pp. 1–10. DOI: 10.1109/IPDPS.2007.370232.

Maciejewski, R., Rudolph, S., Hafen, R., Abusalah, A., Yakout, M., Ouzzani, M., Cleveland, W. S., Grannis, S. J., and Ebert, D. S. (2010). A visual analytics approach to understanding spatiotemporal hotspots. *IEEE Transactions on Visualization and Computer Graphics,* 16(2), pp. 205–220. DOI: 10.1109/TVCG.2009.100.

Malik, A., Maciejewski, R., Elmqvist, N., Jang, Y., Ebert, D. S., and Huang, W. (2012, October). A correlative analysis process in a visual analytics environment. *In IEEE Conference on Visual Analytics Science and Technology (VAST), 2012,* IEEE. pp. 33–42.

Author Biography

Jean Scholtz is a chief scientist at the Pacific Northwest National Laboratory (PNNL). Dr. Scholtz came to PNNL in 2006 to work on the Visual Analytics Science and Technology (VAST) Challenge for Jim Thomas, as part of the National Visualization and Analytics Center (NVAC). Although she no longer works on the VAST Challenge, she continues to advocate and conduct user-centered evaluations for various programs at the laboratory.

Dr. Scholtz previously worked at the National Institute of Standards and Technology (NIST), where she worked in user-centered evaluations for an Intelligence Advanced Research Projects Agency (IARPA) program, Novel Intelligence from Massive Data (NIMD), and collaborated with PNNL to start up the VAST Challenge. At NIST she was a founder of the Common Industry Format (CIF) project that established a format for software companies to describe their usability testing results. This document is used in requests for information today by companies interested in purchasing software and has become an ISO standard.

Dr. Scholtz also worked on human-robot interaction evaluations while at NIST, helping with both Urban Search and Rescue (USAR) and Explosive Ordinance Disposal (EOD) testbeds. She was a program manager at the Defense Advanced Research Projects Agency (DAPRA) involved in collaboration software, digital libraries, and ubiquitous computing.

In an earlier career, Dr. Scholtz worked at Bell Telephone Laboratory in Murray Hill, NJ, where she helped develop an early time-sharing system, Project Mac, with researchers from MIT. She also worked on early efforts at missile defense on Kwajalein, MI.

Dr. Scholtz has a Ph.D. in computer science from the University of Nebraska at Lincoln, a Master's degree in mathematics from the Steven Institute of Technology in Hoboken, NJ, and a Bachelor's degree in mathematics from the University of Iowa.

Dr. Scholtz has been a member of ACM since 1988, has served on a number of CHI conference committees, and was a member of the SIGCHI Board. She received the CHI lifetime service award in 2014. She served for a number of years on the NASA Human Factors in Space Review Board.

Printed in the United States
by Baker & Taylor Publisher Services